T0191943

Designing a Place Called Home

James Wentling

Designing a Place Called Home

Reordering the Suburbs

Second Edition

 Springer

James Wentling
Philadelphia, PA, USA

ISBN 978-3-319-83857-1   ISBN 978-3-319-47917-0   (eBook)
DOI 10.1007/978-3-319-47917-0

Cover Photo: New Port, a new community built on a redeveloped Navy housing site in Portsmouth, VA. (Photograph courtesy of Lusher Productions).

Printed on acid-free paper

This Springer imprint is published by Springer Nature
The registered company is Springer International Publishing AG
The registered company address is: Gewerbestrasse 11, 6330 Cham, Switzerland

Introduction to the Second Edition

This is a book about the design of suburban housing. Following the Great Recession, some media outlets and academics opined that suburban housing is no longer in demand in the United States. One popular book, entitled *The End of the Suburbs*, explained the reasons for the forthcoming demise of suburbia in great detail. Other predictions were that we would become a nation of renters living in multifamily settings—a view reinforced by a decline in the homeownership rate in recent years.

Rest assured, the suburbs are not dead, and the detached house continues to be the most popular choice for most Americans. At the peak of the housing market, over 80 % of all housing starts were single-family detached homes. Since that time, the total has declined: now about two-thirds of all starts are detached houses. Less than 10 % of the total detached houses built are custom homes or rentals, with the balance constructed by homebuilders in suburban settings.

In terms of ownership, detached homes account for over 80 % of our total housing stock, and the latest community survey by the Census Department indicates that that segment is not declining, but rather growing. Attached townhouses and condominiums comprise only 10 % of owned housing. Consumer surveys point to the fact that the younger household who rents still aspires to eventually own a detached home, though preferably in a walkable neighborhood with access to parks and services.

Given the continued preference for suburban homes, and in light of the disdain for "sprawl," it seems appropriate to reconsider the topics discussed in the first edition of this book. For the most part, many of the ideas put forth in the first edition have gained traction. Thanks to advocacy from organizations such as the Congress for New Urbanism, the Urban Land Institute, the American Planning Association, the U.S. Green Building Council, among other groups, new suburban communities have become much more energy efficient, sustainably built, and socially engaging. Many local governments have embraced innovative planning tools such as form-based zoning, conservation communities, and other policies over traditional zoning ordinances that separate uses and encourage automobile usage. Other environmental goals such as reducing carbon emissions, better stormwater and waste management strategies, and promoting healthy mobility options such as biking have also been adopted as policies.

Challenges remain, however, particularly in the area of affordable and workforce housing. And in spite of the progress that has been achieved over the past two decades, the majority of new communities are still built as single-use, detached houses isolated from quality open space and community amenities. More work is still needed to continue the transition toward housing and community design standards that works for all citizens.

With this in mind, and based on an additional 20 years of experience as an architect immersed in residential design and planning, I have edited the first edition of this book to reflect the positives that have taken place, as well as the challenges still to be addressed. Sustainability and energy efficient design practices are among the most important new topics, while demographics and lifestyle preferences are still major influencers on housing and community design practices.

These trends continue to evolve, reflected by floor plans that favor open and casual layouts over formal room definitions. New technologies for windows and doors that allow more natural light into the home without permitting as much heat to enter or escape are more readily available. New heating and ventilation systems are being rolled out to reduce energy consumption and provide healthier interior environments. Outdoor spaces adjacent to the home are becoming more of an extension of the home's interior for easy indoor–outdoor living. New low-maintenance materials for both interior and exterior use are being made from more sustainable materials, such as recycled plastics and textiles, and wood from renewable forestry practices.

Although new technologies have been introduced to new home construction, we find that traditional exterior design concepts for both housing styles and community design continue to resonate with our population. Historically influenced and vernacular house styles with character such as the Craftsman, Dutch Colonial, Mission, and Monterey styles that are familiar to buyers are generally still preferred, even as the Mid-Century Modern style, with its clean lines, is an instance of alternative design becoming very popular.

We will continue to explore manufactured housing in this edition, since it is still a significant portion of owned housing after single family detached. Multifamily housing types will also be included, as they often provide the bulk of our affordable housing stock. Attached housing, for sale or rental, is commonly the initial housing type where younger people reside, as well as for retirees or households looking to downsize or escape the maintenance of a detached house.

Suburban housing and community design can be an extensive topic; to be clear in this book we will be looking at densities of 4–8 units per acre, which includes conventional and small-lot detached homes as well as attached townhouses and garden units up to three stories. We will not be exploring some of the higher density multifamily designs above three stories, a category that generally includes elevator buildings. Nor will there be discussions on specialty housing types such as live/work, co-housing, or accessory units. This book will cover housing that is generally privately built and with designs that could be affordable to buy or rent for households earning the US median household income level. Households with income below that level, including the homeless, are an entirely different topic that generally needs to be addressed by government, nonprofit housing providers, or public-private partnerships—a need for which I have great concern but much less knowledge and experience. Instead, we will focus on the "plain vanilla," the most common and numerous types of housing and communities underway throughout the country.

Finally, environmentally responsible design programs such as Energy Star, LEED, Passive Houses, and others that meet green certification programs will be discussed. These designs represent a growing segment of our new housing stock, and certainly many of the elements of these programs will become the norm in market-rate housing as well as mandated by building codes. We will explore all of these trends and many others in detail in this second edition of *Designing a Place Called Home*.

Acknowledgements

Since this second edition is largely based on the framework established by the first edition, I would like to again recognize those individuals who reviewed and provided guidance on the first manuscript drafted over 20 years ago. Those would include Susan Bradford and Mitch Rouda, then of BUILDER magazine, Todd Zimmerman of Zimmerman/Volk Associates, and James Timberlake of KieranTimberlake Architects. Randall Arendt, then of the Natural Lands Trust, now Greener Prospects, Chuck Graham of Newton-Graham, author Philip Langdon, the late Donald Prowler, professor of Architecture at the University of Pennsylvania, Joseph Duckworth, then of Realen Homes, now Arcadia Land Company, also critically reviewed the text. Rebecca Hardin of Open Line Communications professionally reviewed and copyedited the first edition.

Many of the exhibits and photographs from the first edition were replaced with color images in this publication. Assistance with these images came from photographer Tom Voss, Stephen Fuller of Stephen Fuller Designs, Sara Davis of the Fredrick Law Olmsted National Historic Site, Wiley Books for the National Trust for Historic Preservation, and photographer Rick McNees, who provided the image of the Usonian Home.

Jason Ebberts of TBL Photography provided photos of the Country Club Plaza's historic community and Roger Lewis, professor of Architecture at the University of Maryland and columnist for the *Washington Post*, authorized the use of his cartoons to add some humor to the text. Xavier Iglesias of Duany/Plater-Zyberk Architects, Carson Looney and Beth Van Der Jagt of Looney Ricks Kiss Architects, Mitali Ganguly of Calthorpe Associates, and Debra Ehnstrom and Tony Green of Pinehills, LLC also provided images and background information on their respective communities.

Victor Mirontschuk and Trina Locklear of EDI International, Craig Morrison of Cimarron Homes, Bob Fusari Sr. of Real Estate Service of Connecticut, Dan Mummey of Clear Springs Development Company, Anne Hutchinson of the Natural Lands Trust, Jonathan Braun of Ernest Braun Photography provided the Eichler home photo, and Kristi Jue of Stantec San Francisco (formerly Anshen & Allen) also provided assistance. Jeff Berkus and Ben Krinzman of Jeffrey Berkus Architects provided images from the Berkus Group Architects archives.

Thanks also to Heather McCune of Bassenian/Lagoni Architects, Roxanne Williams and J. Lee Glenn of RLD2X, Holly Stuut and Wayne Visbeen of Visbeen Architects, Donald MacDonald of MacDonald Architects, Jerry Hoffman of RGL Properties, Randall Arendt of Greener Prospects, Kathy West Studios of Village Homes, and Debbie Loudon and Ross Chapin of Ross Chapin Architects also provided images for the text. Finally, Eliza MacLean assisted with images from the *Visualizing Density* book, featuring aerial photography of Alex MacLean and Doug Scott in Seattle provided photos of the High Point community for Mithun Architects.

Marissa Zhao from our practice had the talent and skills needed to add color to the original black and white graphics. She also was responsible for formatting the text and new material from our internal files, as well as for obtaining the photos and graphics from other sources and coordinating the most all of permissions needed for publication of the book.

From Springer International Publishing, Tiffany Gasbarrini was my original editor who accepted the book proposal and then located a "digital brontosaurus" to convert the 20-year-old text from a "floppy disk" (and these were the really floppy ones) into currently usable software. Springer staff members Rebecca Hytowitz, Zoe Kennedy, and Faith Pilacik also provided guidance during the editing process.

As with the first edition, I would like to thank the many clients of our architectural practice, who, in addition to financial support, have supported the firm's ideas by providing built examples of our designs. Many of the ideas in the book were formulated as a joint effort with our clients or by observing the practices of other progressive homebuilders.

In addition I'd like to thank the many talented people who assisted in the formulation and execution of the firm's ideas, the many dedicated employees who have passed though our office, which after 30 years are too numerous to mention by name. A note of gratitude is in order to everyone in this group.

Finally and most importantly, I thank my wife Anne and son Patrick for their patience, love, and support of my professional endeavors.

How to Use This Book

This book is organized from the old to the new, from the large to the small, and from the general to the specific. Depending on the reader's background, some of the chapters may be of more interest than others. Builders of a single house, for example, may have less interest in how the entire community could be better planned. Community developers, on the other hand, might not be as concerned with interior design issues.

1. **Housing Yesterday** is a condensed historical review of planning and design themes found in American housing, with an eye to how some older building practices could be applied to current housing designs.
2. **Housing Today** looks at the forces behind today's housing and community design standards—with suggestions to change them.
3. **Community Planning and Design** addresses land planning techniques to improve neighborhood quality among groupings of houses and housing types.
4. **Siting and Lot Patterns** talks about how to configure houses on appropriate lot sizes—and to organize plans that relate best to the lot and the greater neighborhood.
5. **Floor Plans and Building Image** suggests organizing principles for floor plans to create enjoyable interior room arrangements that will also achieve pleasing elevations.
6. **Interior Details** moves the discussion inside to present more detailed ideas for rooms and spaces that will add warmth, character, and livability with cost-effective details.
7. **Exterior Details** moves back outside the house to recommend design concepts that will enhance exterior elevations.
8. **Multifamily Housing** discusses how general design principles apply to attached houses and addresses community design issues specific to attached houses.
9. **Manufactured Housing** is a brief overview of design concerns of particular interest to builders of manufactured houses.
10. **Toward More Sustainable Homes and Communities** calls for environmentally responsible design practices in new homes and communities, with a particular emphasis on reducing the consumption of natural nonrenewable resources.

December 17, 2002
Weything

Did you ever really consider the wonderful difference in the meaning of the words house and home? A house is a structure to live in. Home—the dearest place on earth—is that structure that is a part of you—made so by its association with your family, their joys and sorrows, their hopes, aspirations and fears. It is a refuge from the trials and struggles of the outer world. It is a visible expression of yourself, your tastes and character.

From Aladdin's 1919 Catalog of Readi-Cut Homes

Contents

Chapter 1
Housing Yesterday

It seems amazing that in our contemporary society most of our new house designs are based on models imported from Europe, some of them over 300 years ago. Take the Cape Cod-style house, for example, which has been built in New England since around 1680. Essentially, a similar design is still built today in housing communities throughout the country. Why? Because the Cape Cod design is ideally suited to deliver maximum square footage within a minimum exterior building envelope.

Other New England prototypes such as the Garrison and Saltbox are also represented in communities in the Northeast, just as the Spanish Mission style remains popular throughout the Southwest. Probably, the single most important event in terms of influencing housing design in the Mid-Atlantic states was the John D. Rockefeller-sponsored restoration of Williamsburg, Virginia, begun in 1926, which put the Georgian architecture of the Chesapeake Bay Colonies in front of millions of future homeowners each year.[1]

This is not a book about the history of housing in America. There already is substantial literature addressing that topic in a comprehensive fashion, many titles of which are listed in the bibliography. I would, however, like to list some historical attributes of American housing in order to identify the characteristics that seem relevant to a discussion of new home and community design. My summary observations would include the following:

1. *American housing designs are diverse in character.*
2. *American housing is typically represented by freestanding dwellings.*
3. *American housing design is most often based on traditional styles.*
4. *American housing is built primarily in moderate density settings.*

Fig. 1.1 Many of our housing designs are often based on models imported from Europe over 300 years ago. The Cape Cod-style house, based on vernacular English cottages, has been built in New England since 1680.

© Springer International Publishing AG 2017
J. Wentling, *Designing a Place Called Home*,
DOI 10.1007/978-3-319-47917-0_1

The above consumer-driven features about America's housing stock have often been challenged. Over the years, many innovative housing types have been introduced, ranging from raised ranches to split-level to postmodern designs that may have been popular for a short period of time but never caught on with housing consumers. While the clean lines of modernism appeals to some upscale households, mainstream homebuyers are attracted to designs that are familiar to them, with details that give their home an individual character.

Earliest Homes: Regionally Diverse Dwellings

Although the earliest American homes, those of the Native American Indians, may not have much application to contemporary production housing design, it is nevertheless interesting to note that they did express the origins of regional diversity. The dwellings of these tribes varied with the climatic and physical characteristics of the regions where they were built. For example, the lightweight, portable Tepee structures built by the Great Plains Indians contrast with the solid, implanted Pueblo structures constructed in the Southwest, and from the earth lodge homes built by tribes on the Northwestern Prairies. Religious beliefs, hunting techniques, and other customs also influenced how American Indian designs evolved.[2]

Since the earliest of times, America has never adopted a national style. Instead, we have a tapestry of regionally diverse architecture. In response to our vast geography, housing designs vary from north to south and from east to west. You may hear people refer to certain housing styles as being "French" or "English." By contrast, there is no one, single American style. Our new homes should and do celebrate this diversity with a wealth of different housing types. In spite of the influence of national organizations, the media and networking technologies that help shrink distances between people and the sharing of ideas, the best new home designs still reflect the social and environmental diversity found in our country.

Fig. 1.2 (**a** and **b**) The restoration of Williamsburg, VA, begun in 1926, put the Georgian architecture of the Chesapeake Bay Colonies in front of millions of future homebuyers each year, sparking a regional demand for that style in the mid-Atlantic states

For the most part, early European settlers would build primarily freestanding dwellings for a single household. The dominance of detached houses in our total housing stock was, and continues to be, unique in the modern world; in most developed countries, government-sponsored multifamily housing, shared housing, or communal living quarters are more the norm than to accommodate mass housing requirements. America, however, settled by people seeking religious freedom and land reform, wanted the opportunity to own property upon which they could farm, build, or own a home. In time, our abundant land and skilled carpenters gradually fashioned a democratic type of housing stock—housing which was more responsive to how people wanted to live, as opposed to what a central government decided was needed.

Although colonial settlers built neighborhoods of handsome row houses in early American cities, and as much as one-third of our new housing is attached in some way, the majority of American houses have always been detached. Almost nine out of ten homeowners and one out of three renters live in single-family dwellings. 83 % of our total housing stock is in single unit structures. When combined with the 7 % of our housing that is manufactured or mobile homes, 90 % of American's housing is detached.[3] This preference for detached homes continues to run deep in the mindset of our population. Homebuyer polls confirm that between 70 and 90 % of potential homebuyers would prefer to live in single family detached homes over attached dwellings.[4] In fact, there is evidence that the preference for ground level, independent dwellings is a universal human behavior.[5]

European Settlers: Simplicity with Style

The earliest colonial homes in America were simple and basic, with the primary objective being survival and protection from the elements. Therefore, most designs were composed of simple forms, mainly rectangular shapes designed to be constructed and inhabited quickly. Early examples include the log cabins of the Delaware Valley, simply defined by four walls, a pitched roof, a door, windows, and a chimney. While materials and building

Fig. 1.3 (**a**) The basic elements of shelter are found in these replicas of the primitive encampments of Washington's troops in the Delaware Valley. (**b**) Drawings by children show how these shapes are almost universally associated with domestic architecture in America. (Figure (**b**) courtesy Cimarron Homes)

technology have changed over the years, this basic geometry would continue to define all but a fraction of American homes built for the next three centuries. Furthermore, these elements and shapes are almost universally associated with the idea of domestic architecture in America—just look at any child's drawing of a house and see!

Houses with flat roofs, geodesic domes, and other futuristic concepts are typically not found in suburban communities. Americans have preferred to live in dwellings that are traditionally styled and familiar to them. That is why homebuyer polls reveal an overwhelming preference for historical, traditional styles over contemporary architecture.[6]

Some of the earliest homes to express stylistic influences were the Dutch Colonial homes of the Hudson River Valley in upstate New York. As early as 1650, the Dutch built housing with symmetrical plans, articulated gables, exaggerated roof projections, entry porches, and dormer windows. Many of these architectural elements remained popular through the twentieth century, including the split "Dutch door" originally designed to allow natural breezes to enter the house while simultaneously keeping barnyard animals out.

Settlers in California built what would become know as the Mission-style house. Unlike other forms of colonial housing, the original Missions were based on communal living as they accommodated both religious and military personnel in a unified structure. The predominant plan was a single-level courtyard configuration, with rooms connected by outdoor protected porches. This plan was particularly well suited to the region's Mediterranean climate, and the whitewashed adobe walls and red barrel clay tile roofs of the mission style are still dominant materials in most regions of California and throughout the southwest. Further, the single-level ranch plan continues to dominate the entire southwestern housing market, with many houses now incorporating new variations of the courtyard themes.

Another colonial style that developed during the latter part of the seventieth century was the Southern Colonial homes of the Southeast, with their south-facing porches. Typically incorporated on the front of the house, porches allowed outdoor living during hot summer months. Front porches would later become a stylistic icon of American housing throughout the country, symbolizing hospitality and connection to the greater neighborhood.

Fig. 1.4 In the Southwestern part of the United States, the traditional Spanish motif remains popular, as seen with this modest house in Coronado, CA. (Photograph courtesy Tom Voss)

Fig. 1.5 (a) Washington's *Mount Vernon*, built 1735–1785, with its two-story portico, would be emulated by residential builders for the next two centuries, as seen here, in a rural North Carolinian residence (b)

Agricultural Age: Jefferson's Low-Density Arcadia

In the 1700s, colonial America matured into an agrarian nation of family farms and large estates, such as George Washington's *Mount Vernon*, and Thomas Jefferson's *Monticello*. Gentlemen farms such as *Mount Vernon* and *Monticello* were part of Jefferson's ideal for the development of the continent as a low-density collection of farms and hamlets. His vision for the United States as a nation of freeholders contrasted sharply with the feudal land ownership that dominated Europe at the time, a system the new Constitution sought to prevent.

Mount Vernon, built over a 50-year period between 1735 and 1785, would be emulated by residential builders across America for the next two centuries. The modified Palladian architecture found at *Mount Vernon* is still considered by many current homebuyers to represent an ideal image for a prestigious family residence. Design themes include the entry façade, which is a composition of smaller service wings enhancing a larger and more imposing main house, and which collectively define a gracious entry courtyard. Other themes from *Mount Vernon* include the nearly symmetrical placement of windows, doors, dormers, and chimneys, and the grand-scaled elements such as the high-columned piazza, extending the full length of the East Front overlooking the Potomac. These details from *Mount Vernon* established precedents that would be replicated time and time again in upscale American homes.

Jefferson's *Monticello*, sited on a mountain in the Blue Ridge range of Virginia, overlooked a countryside of privately owned farms sustaining mercantile villages. The family farm, with its collection of buildings, and the farm village, would become models for later American housing and community developments. While the origins of the modern American suburb are generally traced to the expansion of industrial London, the agricultural villages and farms of colonial America would be another theme for future settlement patterns across the continent.

Estates such as *Mount Vernon* and *Monticello*, along with the more modest family farms, represented the American Dream to own land—land upon which a home could be sited. This ideal vision of the home is undeniably part of the market forces shaping the planning and design of new home communities today.

Fig. 1.6 (**a**) Jefferson's *Monticello* (**b**) overlooked a landscape of family farms and villages (**c**) that would evolve into moderate density settlements—forms which are emulated in new home communities today

a

b

Fig. 1.7 (**a**) This illustration of a Romantically styled house from A.J. Downing's <u>Victorian Cottage Residences</u>, published in 1842, is similar to the images people are drawn to today, as seen in (**b**) Stephen Fuller's rendering of an idea home for a building supply company. (Figure (**b**) courtesy Stephen Fuller Designs)

Industrial Age: Stylistic Diversity

As early agrarian settlements matured into larger towns and cities, residential architecture lost some of its regional flavor. By the 1840s, the industrial revolution was dramatically changing how people perceived and built their homes, as seen by a marked increase in various new residential styles. For example, the Romantic houses which included the Greek and Gothic Revival movements became increasingly popular during the first half of the nineteenth century, partially in response to a machine-age society that made people yearn for less "industrial looking" houses.

The new Romantic styles were able to proliferate in part due to the widespread use of pattern books that provided homebuilders with plans and elevations being used throughout the country. The most well known of these pattern books was Andrew Jackson Downing's <u>Victorian Cottage Residences</u>, published in 1842. Through the use of pattern books, housing design was brought into a world similar to fashion, where homebuyers had several options to choose from—and selecting the most appropriate look for one's home was a major decision.

As a result of the use of pattern books, mid-nineteenth century communities became more architecturally diverse. Homebuilders would now erect different styles of houses right next to one another. In the industrial age, it was not unusual to find an Italianate house next to a Greek Revival, and so on, in a fashion that still occurs in new communities today. The use of "patterns" obtained from books and magazines remains a common method for homebuilders to select and establish designs for their new houses.

Jackson's publishing must also be credited with developing the current "homespun" image of suburban housing with his ideas about gardens and villa designs. During the Romantic era, people focused on making their homes look natural and domesticated, and Jackson's dislike of the formal symmetry of most colonial styles was well suited to the national mood. House and garden publications that published designs by Jackson and others became major influences on the interior and exterior decoration of houses. The most popular song of the era was *There's No Place Like*

Fig. 1.8 (**a**) The Greek Revival style of this nineteenth century house in Wickford, RI is emulated in contemporary production houses, such as this model at *Old Farms* in Middletown, CT (**b**)

Home, and Jackson's Romantic designs seemed to convey the feelings of what people wanted their homes to look like.

Another significant characteristic of the Romantic houses was the practice of creating styles primarily through exterior decoration. Beginning with the Romantic homes, *beauty was separated from function*. Floor plans of Romantic houses were not overly articulated; designers would take a basic rectangular or cross-shaped plan and—by simply applying decorative elements to the façade—emulate any number of period styles. Gothic Revival homes, for example, were distinguished largely by windows that consisted of elongated shapes and an occasional pointed arch in the upper sash, accompanied by bracket moldings on roof eaves and porches. Greek Revival homes, by contrast, used restrained decoration on the exterior in deference to the balanced simplicity of classical Greek architecture.

By allowing different styles to travel freely throughout the country, within a neighborhood, and along a street, *the Romantic homes of early industrial America established a precedent for architectural diversity within residential communities*. Second, the Romantic houses began a tradition of expressing styles primarily through *decoration of a basic plan with appliques*. This relatively inexpensive method of attaching smaller, detailed elements to a larger simple frame remains a popular way to add character to modest production houses.

Victorian Era: More Complex Designs and Early Suburbs

Britain's Queen Victoria reigned in England from 1837 to 1901; however, the Victorian era in America is generally thought of being from 1860 to 1900. During this period, industrial America saw more technological innovation in residential construction. For example, the lightweight balloon framing made of 2 × 4 lumber was replacing heavier timber members for structure, allowing floor plans to break out of rigid, box-like shapes and into more complex forms. The rapid expansion of railroads and manufacturing capacity allowed prefabricated building components and even whole houses to be factory made and shipped to the site by rail. As a result, Victorian homes were defined as elaborate stylized dwellings—previously available to only the wealthy, but now made affordable to the masses through the economies of industry.

Fig. 1.9 (**a**) The Victorian style, as seen in this stick-style house in Eureka, CA, became popular during the Great Western Expansion of the late nineteenth century, and it has remained a popular style today, as seen in these infill houses in downtown Charlotte, NC, built by JCB Urban Company (**b**)

The use of Victorian architecture coincided with the Great Western Expansion of the continent, and it is still a dominant historical style in Western cities such as San Francisco and Denver. During the Victorian period, floor plans began sporting bay windows, conical towers, and other whimsical projections. Rooflines were varied, and intricate wraparound porches became signature of Victorian architecture. Sophisticated material production during the Victorian era allowed a myriad of patterns, colors, and shapes to be used on roofs and walls. Elaborate trim and molding details appeared on roof cornices. The Victorian era ushered in building plans and details that are now difficult to produce within production home budgets.

During the Victorian period, early suburban communities began to take shape. The foundations of today's suburbs can be traced back to residential communities that were once organized according to the means of transportation that allowed them to occur. The earliest examples were made possible by railroads, then light rail trollies, and finally, highways and freeways. Railroad suburbs included Chestnut Hill outside Philadelphia, dating from 1854, and Riverside, Illinois, started in 1869.

Riverside, master planned by the firm of Olmsted, Vaux and Company on 1600 acres nine miles from the business center of Chicago, was organized in a hierarchal fashion. Since residents commuted to the city, the master plan identified the Burlington and Quincy railroad station as a focus, with commercial services and higher density housing around it. Lower density large-lot single houses were sited along curvilinear streets connecting to the station and commercial center.

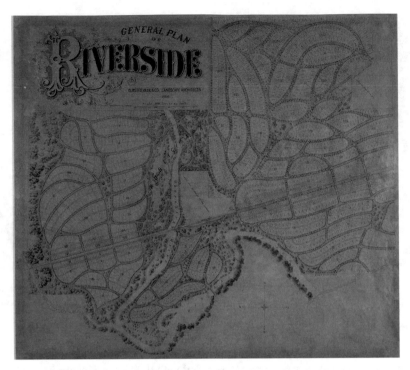

Fig. 1.10 The curvilinear street pattern of the Illinois railroad suburb *Riverside* was master planned by Olmsted and Vaux on 1600 acres, nine miles west of Chicago, and was started in 1869. (Courtesy of the United States Department of the Interior, National Park Service, Frederick Law Olmsted Historic Site)

Fig. 1.11 *Riverside*'s village pattern of mixed-use and mixed-density housing with extensive open space would become a model for subsequent trolley and automobile suburbs until the postwar era. The signature water tower is seen here adjacent to the rail station. (Photo courtesy the Village of Riverside)

Riverside was bisected by the Des Plaines River, and Olmsted devoted 700 of the 1600 acres to common space, including a 160-acre park. Residential lots were no less than 100 ft wide by 200 ft deep. In addition to the residential sites, the Riverside Improvement Company built a stone chapel with seating for 300, a block of stores and offices with a drug store, market, post office, and other suppliers. *Riverside* also had a hotel with 124 rooms, dining rooms, and gardens. A private school was built in 1874.

Most of the early houses at *Riverside* were individually designed by architects in Romantic or Victorian styles; Riverside also has some of Frank Lloyd Wright's most famous Prairie-style houses within its borders. The Romantic architectural styling of the public buildings achieved the rural village character that Olmsted and Vaux sought to establish. This rural character was further enhanced by preserving the wooded lands of the site and creating new natural settings throughout the plan.

The suburban village pattern seen at *Riverside*, with mixed-use amenities and mixed-density housing, was also the rule in planning subsequent trolley and automobile suburbs up until the postwar era. However, as we will discuss later, planners now advocate a return to these models in today's new communities, with *Transit-Oriented Development* promoted as a community planning approach to encourage transit use and reduce automobile dependence.

Eclectic Houses and Modernism

During the period from 1890 to 1920, more houses were built than in the Nation's entire previous history.[7] There was also an unprecedented diversity in housing styles. The early twentieth century saw the introduction of the Bungalow, Four-Square, Mission, and Tudor styles of the *comfortable house era*, many of which are still lived in today. For the bulk of this post-Victorian period, there was an optimism in society that had become reflected in the design of residential architecture.

Fig. 1.12 (a) The Four-Square house as seen in Denver, CO, and (b) the Tudor-style house as seen in Longmeadow, MA were both popular during the early twentieth century and can still be seen almost nationwide

Fig. 1.13 (**a**) The Craftsman style, found in this 1920s house in Greensboro, NC, is still popular today. We see this style emulated in an entry-level design at *Falls River* in Raleigh, NC (**b**)

After the turn of the century, further advances in building technology gave residential designers even greater latitude to emulate traditional styles. Lightweight wood framing could accept applied veneers of brick or stone, enabling architects to replicate very old European masonry buildings, such as those of the French Chateauesque and English Cotswold styles. This eclectic period-based trend of replicating historical European styles with new building technology would continue into the next century, but not without criticism from advocates of the modern movement.

Modernism's influence on the American suburban landscape was primarily through the still-popular Craftsman style, and the closely aligned Prairie style advocated by Frank Lloyd Wright. *Proponents of the craftsman style felt that if we must live in a machine age, houses should present an image of craftsmanship and the art of building by hand.* Although most craftsman homes were built from machine-tooled components, the style embodied the spirit of *craftsmanship*, *domesticity*, and *human scale*, all of which are design qualities still considered desirable by most of today's homebuyers.

Fig. 1.14 Sears, Roebuck & Company's mail-order houses, sold from 1908 to 1940, used traditional styles such as the Dutch Colonial. Shipped to the site by rail, Sears houses can still be seen throughout the country. (**b**) The Dutch Colonial style is emulated today at *Main Street*, a new home community in Metuchen, NJ. (Figure (**a**) copyright Sears Catalog Archives. Reprinted with permission)

Fig. 1.15 The Usonian style developed by Frank Lloyd Wright is seen in his design for the *Jacobs First House* in Madison, WI, built in 1937. Elements of the Usonian style are found today in contemporary ranch houses nationwide. (Photography courtesy Rick McNees)

Craftsman homes and homes in other revival styles introduced in the eclectic period are still very popular, with favorite styles including the Colonial Revival, Spanish Revival, and French Eclectic. Additionally, Frank Lloyd Wright's Prairie style and Usonian homes influenced later ranch and contemporary designs. All of these styles were adapted by homebuilders to create a sense of perceived value for their products in the American homebuyers' eyes.

Eclectic house styles proliferated throughout the country with the introduction of *mail-order houses* offered by companies such as Sears and Roebuck, Aladdin, and Honor-Bilt. From 1908 and 1940, Sears and Roebuck sold approximately 100,000 homes with approximately 450 ready-to-go designs that were shipped by rail directly to building sites.[8] These prefabricated designs impacted American suburban communities in a fashion similar to the pattern books: they allowed styles to travel freely throughout the country. As a result, one can roam neighborhoods in St. Louis and see the same house observed earlier in Maryland or Georgia.

Trolley and Early Automobile Suburbs

The first quarter of the twentieth century saw the early planned streetcar and automobile suburbs, epitomized by places like Forest Hills, New York. Planned by the Olmsted brothers and built between 1909 and 1912, Forest Hills followed the model of Riverside, Illinois which Olmsted and Vaux had previously designed with mixed-use and mixed-density housing. Architecturally, Forest Hills was

Fig. 1.16 Forest Hills Gardens, which was planned by Fredrick Law Olmsted, Jr. and begun in 1909, was linked to Manhattan by the Long Island Railroad. Sponsored by the Russell Sage Foundation, and influenced by the Garden Cities Movement, it was intended to be a model suburban community for the automobile age. Architect and housing reformer Grovsner Atterberry based his designs on the English Cotswold style

built in an English country style to emulate a Romantic village atmosphere. Other trolley suburbs of the day included *Shaker Heights* outside Cleveland, Ohio from 1916 and *Roland Park* outside Baltimore, Maryland, started in 1891.

Early automobile suburbs included Beverly Hills, California, started in 1906, and *River Oaks* outside Houston, begun in 1923. Perhaps the best model of an automobile suburb, however, is the *Country Club District* in Kansas City, Missouri (and Kansas), begun in 1907 by developer Jesse Clyde Nichols. Starting with a modest 10-acre tract, the J.C. Nichols Company would over 30 years expand the District to over 4000 acres including 6000 homes, 160 apartment buildings and seven shopping centers—home to over 35,000 people.[9]

Master planned by George Kessler based on the City Beautiful Movement, the *Country Club District* included generous parks and open space. The centerpiece of the district was *Country Club Plaza*, a shopping center in a Spanish Village motif with shops organized around squares and interspersed parking. Multifamily housing of various densities surround the 15-block shopping center. There are also four golf courses, 11 churches, and 15 schools in the Country Club District.[10]

Fig. 1.17 The centerpiece of the Country Club District was the Country Club Plaza, a shopping center modeled after a Spanish village. As part of the design motif, the Plaza's *Giralda Tower* is a replica of the original in Seville, Spain. (Courtesy Jason Ebberts/TBL Photography)

Because the District's land was outside the central city, Nichol's strategy was to provide a residential setting of such high quality that buyers would be enticed to travel by car to live there. In doing so, the Company developed a community design philosophy worth recounting today. In a 1939 interview, Nichols revealed a few of his design goals: *"the use of curving streets to fit the contours of the land," "small triangular parks at street intersections ... generously planted with shrubbery and trees to give a park effect,"* and *"after you establish a new neighborhood, it is important to set up a homes association and develop other neighborhood activities, so as to maintain neighborhood morale."*[11]

While the Country Club District represented a privately sponsored new community for the middle class, other modest examples of early suburbs include the *factory suburbs*, built by corporations to house workers and their families. These included Pullman, Illinois, started in 1880 (Pullman made railroad cars) and the Olmsted-planned town of Kohler (a plumbing manufacturer) in Wisconsin, in 1909. Factory suburbs differed from transportation suburbs in that residents lived *near* their place of work and housing was subsidized by the company sponsor. Today, employers are still involved with helping to provide housing for their staff, a practice that is particularly popular with colleges and universities.

Fig. 1.18 This house in Levittown, NY was one of six basic models used for a community of 17,447 households. The selection of the Cape Cod style reinforces the style's enduring appeal and frugal qualities

Information Age: Postwar Building Boom and the Automobile

Homebuilding slowed dramatically in the 1930s depression period and then virtually ceased during World War II. Once construction resumed in 1946, there was a renewed sense of urgency to generate massive amounts of housing to accommodate returning GIs—as well as to make up for the past several decades of minimal production. This postwar building boom, supported by government finance programs, would continue for the next 40 years.

As America became an information-based society, settlement patterns changed dramatically. In 1956, President Eisenhower signed the Federal Aid Highway Act that authorized construction of the 42,500-mile Interstate Highway System, sparking the decentralization of cities. Massive "bedroom communities" where workers could buy homes at affordable prices within driving distance of employment emerged as the postwar standard for new housing.

Residential design in the 1950s was characterized by generally straightforward plans with varying degrees of regional- or traditional-applied ornament. The simplified ranch and modified Cape Cod houses were immensely popular. New plans that debuted in the 1950s included the split-level and bi-level designs—generally derivations of plans from the modern movement that offered internal zoning for family activities as well as economy of construction.

Perhaps the most quintessential postwar community was *Levittown*, built by developer William Levitt on Long Island, New York. After gaining experience with mass-produced housing during World War II, Levitt and Sons was poised to serve the needs of returning GI's. Working in assembly-line fashion, the company began development of *Levittown* in 1949 with a 700-square-foot model house on a 60 by 100 ft lot, which sold for $7990. On opening day, Levitt took contracts for 1400 houses and completed the 17,447-house community by 1951.[12] Although Levitt originally refused to sell houses to black buyers, today the Levittown in New York, Pennsylvania, and New Jersey now provide affordable housing to racially and ethnically diverse populations.

Although the Levittown communities were often derided for sameness of design, master plans included parklands, school sites, convenience shopping, and even public swimming pools, making them examples for how to deliver affordable housing with neighborhood amenities in the present day. Architecturally, one of the base *Levittown* houses was a modified Cape Cod, which again reinforces the appeal of traditional design and the frugal, yet enduring qualities of the Cape Cod design.

Postwar communities and individual houses had, however, gradually been designed more in response to the automobile and less toward human and social concerns. Furthermore, as new highways allowed the decentralization of cities to accelerate, planning boards in rural communities were pressured to adopt the low-density "slow-growth or no-growth" zoning policies that eventually would become quite devastating to community design standards in the twentieth century.

I think the poor design that started in the 1950s is an appropriate point for ending this discussion of the housing of yesterday, and for beginning to look at the housing of today. Before leaving the historical context of American housing, however, I would like to restate some of the design characteristics that seem applicable to new production homes:

1. *American houses are regionally diverse in character*, reflecting the varied climate, landscape, and cultural traditions within our country. America does not have a national residential style, but rather a tapestry of regionalized design, and our best residential designs reflect our social and environmental diversity.

2. *American housing is typically represented by freestanding dwellings*, surrounded by open space, accommodating a single household. America's private homebuilders fashioned a housing stock responsive to what people wanted, rather than government imposed solutions. Because of this, roughly 80 % of our housing stock is detached.

3. *American housing is most often based on traditional styles* imported from Europe. Early styles were simplistic. Later, the complex styles of the Romantic and Victorian periods were emulated with applied ornamentation on basic frames. Designs were taken from pattern books and mail-order companies, and consequently, different period styles were frequently mixed within a community.

4. *American housing was built in moderate density settings*, initially defined by agricultural land uses, and later evolving into medium-density suburban communities around established cities. Suburban community design varied with its relationships with employment and transportation. Early suburban communities were defined by mixed-use and mixed-density development, with integral commercial and institutional services.

In the next chapter, I will begin a discussion of the specific design problems and concerns found in postwar communities—problems that may be addressed by a reconsideration of some historical design themes and settlement patterns from yesterday.

Notes

1. Stern, Robert A.M. 1986. *Pride of Place, Building the American Dream.* Boston, MA: Houghton Mifflin Company. p. 331.
2. Walker, Lester. 1997. *American Shelter Revised Edition: An Illustrated Encyclopedia of the American Home.* Woodstock, NY: The Overlook Press. pp. 22–37.
3. U.S. Census. "2013 Housing Profile: United States." *American Housing Survey Factsheets.* May 2015.
4. Quint, Rose. "Housing Preferences across Generations (Part I)." *Eye on Housing.* National Association of Home Builders: 12 April 2016.
5. Turner, John F.C. "Toward Autonomy in Building Environments." *Housing by People.* 1991.
6. Miller, Meredith. "Home architecture style: Regional or not?" *Zillow Research.* 12 March 2013. <http://www.zillow.com/research/home-architecture-style-regional-or-not-4388/>
7. Gowans, Alan. 1986. The Comfortable House, North American Suburban Architecture 1890–1930. Cambridge, MA: MIT Press. p. xiv.
8. Stevenson, Katherine Cole and H. Ward Jandl. 1986. *Houses by Mail.* Washington, DC: The Preservation Press. p. 30.
9. National Real Estate Journal, February 1939, "The Country Club District, Kansas City." p. 20.
10. Ibid. pp. 26–33.
11. Ibid. pp.
12. "Levittown: the Archetype for Suburban Development." *American History Magazine.* October 2007.

Chapter 2
Housing Today

In this chapter, we will discuss the challenges facing builders and developers attempting to build (as well as homebuyers attempting to buy) an attractive, afford-able, and comfortable new home in a cohesive neighborhood. As we discussed in the previous chapter, postwar communities were increasingly designed to accom-modate the automobile, both in terms of their streets and their houses. Other forces from planning and zoning controls, neighborhood organizations, increased land costs, and financing requirements all contributed to the development of stereotypi-cal low-density communities that have come to be known as "suburban sprawl."

Since the first edition of this book was published, there has been a tremendous improvement in new community designs found in many areas of the country, pri-marily thanks to the Smart Growth and New Urbanism movements that we will discuss later in this chapter. However, in my travels I still see the same-old stereo-typical, tired designs way too often, which indicates there is more work to be done. This is perhaps more true in well-established regions outside legacy cities, where change comes slowly and resistance to new ideas is common.

Too many of these recent communities, or subdivisions, are still known for excessively wide streets without sidewalks. In the East or Midwest, the homes are typically spaced apart on large lots that range from one-quarter to two acres in size. Landscaping tends to be sparse, and common open space for parks or playgrounds are rarely included within the community.

The homes themselves tend to be box like in shape with the garage set off to one side or in front of the house. The walkway to the front door comes off the driveway since there is no sidewalk in front of the house. The homes are generally all one

Fig. 2.1 Excessively wide streets, oversized lots, and box-like houses are still woeful realities of new home communities today. This is a common scene in the Eastern and Midwestern parts of the country

© Springer International Publishing AG 2017
J. Wentling, *Designing a Place Called Home*,
DOI 10.1007/978-3-319-47917-0_2

Fig. 2.2 In the Sunbelt and Western states, communities can look like this: tiny lots with narrow frontages, oversized houses dominated by garage doors, and little sign of habitation when viewed from the street

material—siding or stucco—with the variation between houses being primarily in color. Some homes may incorporate an accent material, often brick or stone, which is used only on the front façade to keep building costs down.

The homes that make up the community may include several models distinguished by changes in rooflines, plan shapes, window and door placements, and other elements. Often the window sizes and locations are awkward though perhaps related to some interior layout. Other variations in the façades seem to lack a sense of balance.

In higher density communities in the West or parts of the South, there are other problems. Lots in these locales can be very small, sometimes as little as 40 ft wide by 90 ft deep. Here, the street view of the homes is very poor, and you might even see homes where the entire façade seems to consist of garage doors and driveways. In some cases, entry doors are moved back and to the side of the home and are not even visible from the street. There may not even be any windows at all on the street elevation.

Indeed, there are still far too many new home communities that fit this sad description. What follows are some reasons why our neighborhoods evolved from the charming trolley and early automobile traditions described in Chap. 1 to what is often called subdivisions, tract housing, or simply suburban sprawl.

Mass Production and Financing

After World War II, there was a need to build houses rapidly and inexpensively for returning GIs and to make up for the dearth of housing built during the depression, housing construction moved into a factory-like mode of mass production. William Levitt's famous assembly line system of delivering houses can be credited with helping to double the prewar homeownership rate from the mid 30 % to the mid 60 % range, as did similar builders nationwide. This prompted builders to question

Fig. 2.3 In the postwar era, houses were stripped of frills such as covered porches, articulated floor plans, and other details. The goal was to build to the maximum square footage at minimum cost per square foot

whether some design characteristics might be considered "frills," and soon gone from many postwar houses were covered porches, articulated floor plans, quality veneer materials, and interior built-in features. Garages or carports were pulled up to the side or front of the house to reduce construction costs as well.

Savings from these more basic designs were converted into additional interior space. The objective at this point was to deliver the maximum square footage at the least possible cost. Due to the rising number of houses that were financed by mortgages, real estate appraisers often dictated the valuation of homes, and did so primarily based on square footage, not by material use or design quality. Furthermore, the federal government's role in guaranteeing private mortgages through the Federal Housing Administration (FHA) and the Veterans Administration (VA) resulted in houses being measured according to "VA/FHA Standards," which dictated minimum room sizes and material specifications. These minimum standards, published in 1938, did not address community design or exterior design features.

The emphasis on square footage valuation continues to influence housing design today in almost all market segments. The benchmark that realtors tell homebuyers to use to evaluate their options is the relative price per square foot of available models. Thus, homebuilders put their designs up against other builders they compete with in the same market, and often need to consider omitting exterior features that would make their model more attractive in order to keep their price per square foot in line with the competition.

Inside new houses, details and materials have also been downgraded to reduce per-square-foot costs. Quality finishes such as ceramic tile, moldings and trim, wainscots, and turned stair railings, which were once common on even modest homes, are now found primarily in "upscale" houses. Charming details such as built-in shelving, bookcases, fireplace surrounds, and window seats are also increasingly rare. Labor costs, particularly for craftsmanship-intensive labor, gradually made these amenities increasingly difficult to justify in production housing.

The Zoning Game

Local government typically exercises a heavy hand over the design of our communities and individual homes though usually in a negative way. The American system of local government controlling land use decisions has expanded in scope year after year since the early part of the century when zoning first became popular with municipalities. Planning and zoning board members often give in to pressure from citizen groups or individuals who attempt to block any new housing growth in the interest of protecting their own property values. Overzealous control over land use and construction standards by local, state, and federal government agencies is well documented and acknowledged to have driven up the cost of new homes by as much as 24 %.[1]

From a design perspective, the majority of impairment from local government comes from antiquated zoning or subdivision regulations that are sometimes adopted with the ulterior motive of keeping housing costs high. Excessive street right-of-ways and paving, along with oversize lots, large building setbacks, excessive parking requirements, and low-density caps, do little to promote affordability or flexibility in community design.

Fig. 2.4 As architect/cartoonist Roger Lewis reminds us, planning and zoning laws have been controversial since their inception in the early twentieth century. Currently, government controls over land use are acknowledged for driving up housing costs by as much as 24 %. (Courtesy Roger Lewis)

Additionally, planning approvals typically involve "impact fees" for any number of reasons, ranging from street improvements to open space funds to schools, and to affordable housing trust fund contributions. Inclusionary zoning may require that a small percentage of homes be offered at below-market prices for qualified buyers. This sounds admirable, but it drives up the cost of the remaining units meaning that the average price per unit is increased while the percentage of affordable units added to the market is minimal.

Other planning and zoning requirements may dictate very specific design requirements for houses, such as acceptable veneer materials, minimum percentages of windows and doors on certain façades, setback distances for garages, and limitations on garage width and depth. While most of these are well meaning and, in some cases, can help the appearance of homes and neighborhoods, it is questionable whether they represent reasonable uses of the municipalities' "police power" to protect the health, safety, and welfare of its citizens. Moreover, they often tie the hands of builders to formula-like solutions that may or may not be contextually appropriate and, more often than not, increase costs without any real benefits to the buyer.

Although the concept of local control over land is well rooted in American government, the complexity of the task can be technically and politically overwhelming for small communities and their administrative budgets. Many planning board members are appointed in a politically charged atmosphere with either a growth or no-growth mandate—hardly an atmosphere that encourages implementing well-designed communities.

Building Costs Flow from the House to the Land

As the postwar trend toward decentralization gathered momentum, demand for suburban land suitable for residential use began to rise, and landowners in turn increased their prices. Historically, builders liked to keep their finished lot costs to around 20–25 % of the ultimate sales price of the house. Therefore, when land prices went up by 10 %, builders would have needed to raise the price of the finished house by 40 % in order for the land cost to remain at 20–25 % of the house. In other words, if the price of a lot escalated by $5000, the overall home would need to go up by about $20,000.

Gradually, the 20–25 % ratio forced the price of housing beyond acceptable levels of affordability, and builders tried to respond by cutting other costs. Initially, they attempted to compensate for increased land prices by reducing material costs: brick exteriors became siding, tile roofs became asphalt shingles, etc. Prefabricated building components and less expensive interior finishes were other cost-cutting methods that builders embraced in order to stem the escalating price of their product without cutting square footage. The bottom line for consumers, however, was that housing prices were higher, with more dollars devoted to land than to the physical structure.

During the 1970s and 1980s, designers and planners suggested that to mitigate increased land costs, new forms of higher density small-lot housing could be used. This movement started in the West Coast, where land costs were highest, consequently seeing the introduction of prototypes such as the "zero-lot-line" house that eliminated one sideyard from the traditional lot form. Other house/lot configurations, such as "Z-lots," "wide-and-shallows," and "zipper lots," soon followed. However, although these new housing prototypes increased density, in reality this did little to help with the affordability problem, while introducing even lower standards for unit and community design. Thankfully these concepts have for the most part disappeared in the new millennium.

Since raw land is valued according to the development rights conferred by local governments, increased density benefits only the obtainer of those rights. In some cases that may be a homebuilder, or even a builder of affordable housing, but they were the exception rather than the rule. Most homebuilders purchased improved lots at market-rate prices after any benefits for increased density had already been realized by the land developer.

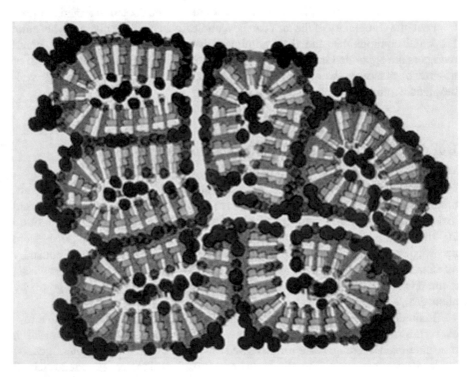

Fig. 2.5 Higher density housing was embraced to reduce land costs during the postwar era. Many of these small-lot housing concepts diminished community design standards and did little short-term financial good

Even when homebuilders bought the land and then obtain increased density rights, the actual shrinking of lot sizes is only significant when the density increase is dramatic, like a 50–100 % increase. A 25 % density increase is insignificant on the ultimate price of a house. The real savings in land development is in site improvements, i.e., reducing road widths, utility design standards, developer exactions, and other entitlement costs. Reduced site improvement costs, however, can be achieved independently of reduced lot sizes.

Increased density and small lots that exceeded rational planning standards (such as floor-area ratios or impervious surface coverage) did little to reduce housing costs, yet still inflicted tremendous damage to the social and physical character of the community. A better way to cut land costs is to convince local government to reduce *excessive* site improvement standards, such as street widths and setbacks. Major savings can also be achieved by accelerating the approval processes and eliminating unreasonable permit and impact fees that drive up the cost of buildable lots.

Home Buyers Focus on Interiors

One of the reasons why the quality of the outside of houses diminished is that many buyers were more concerned about the inside of the home anyway. As noted in Robert Putnam's research documented in his book Bowling Alone, people are increasingly pulling away from group activities, such as block parties and neighborhood picnics, and internalizing themselves within the household structure—much to the detriment of the "social capital" of the community.[2] The increased presence of the internet and home entertainment options has bolstered this trend, keeping people glued to their computer or television screens.

I once heard Ray Watson, former chief planner of Irvine Ranch (now a community of over 80,000 high-density homes in Orange County, California) state that market research indicated potential homebuyers visiting their communities had no expectation that their home would look any different than any other home, and that buyers purchased homes based on the identity established by the image and location of the community only.[3] Home owners at Irvine Ranch had no expectation of owning a home that looked any different from the others around them. Therefore, the only means for them to establish a sense of homeowner identity would be inside the home.

Thus, today's new houses are sold from the inside. Homebuilders recognize the need to have "curb appeal" on the street elevation, but beyond providing an attractive front façade for the model home, design priorities are focused on the inside of the home. When marketing new homes, the objective of decorating one or more model homes with furnishings and accessories is to let the buyer be captivated by the model house. Community location, schools, and access to services and amenities are also part of the marketing strategy, but the interior design is likely one of the most important considerations of prospective buyers.

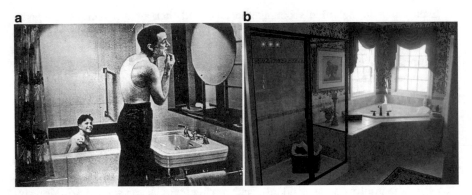

Fig. 2.6 (a) This 1939 advertisement, when contrasted with (b) a contemporary bathroom design, shows how dramatically interior spaces have been upgraded during the postwar years

Designs from Architects and Stock Plans

One of the main paradoxes of the postwar decline of quality design is the corresponding increased involvement of design professionals in the housing industry. Prior to World War II, architects were virtual nonparticipants in the housing field and most residential land planning was done by civil engineers. Homes were typically built from plans that were found in catalogs or pattern books, which were then revised by local drafting shops at the homebuilder's directive.

Starting in the 1960s and 1970s, however, architects in America became more involved with production housing design, particularly on the West Coast. As suburban housing became more dense, some firms saw the opportunity to convince builders that professional design services were needed to make their product marketable. As previously noted, architects and planners focused on higher density detached designs, some of which often were at odds with rational planning and design standards. Some architects also became overly involved with interior space planning concepts instead of looking holistically at integrating interior and exterior features, as well as lot and street orientations that go hand in hand with quality community design standards.

Many new houses still continue to be built without architects, from plans obtained from catalogs and plan books similar to A. J. Downing's pattern books in the 1800s. Since most state and local governments do not require an architectural license to design a single-family home, stock plans are distributed and marketed nationwide very effectively and are now more available than ever due to the internet and web sites that allows easy browsing of house design options.

Stock house plan services provide small builders with the opportunity to purchase what may be a sophisticated, professional design at a fraction of what it would cost to have a custom design developed by an architect. By selling the plans in volume, plan services can recoup their costs to develop new plans with bulk sales.

One significant problem with the stock plan process concerns the matching of designs with lot configurations, local building technologies, and architectural traditions. Does a house that was designed for Houston work in Peoria? While stock plans can be modified to meet the regional context, often they do appear out of sync with the climate, lot configuration, local styling, or community character.

Large-Scale Builders Can Lose Human Scale

Home building in America is big business. The National Association of Home Builders estimates that residential construction and related products manufactured for the industry comprised 15–18 % of the total Gross National Product in 2009.[4] Large national homebuilders, such as D.R. Horton, Lennar, and Pulte Homes, build over 15,000 houses each year in many different markets. In California, large homebuilders consider a small job to be 150 houses, and to be profitable they may need to build them all at one time.

In his 1987 book, The Rise of the Community Builders, Marc Weiss traced the evolving role of twentieth century residential builders, from small craftsmen building individual houses to giant corporations that subdivided large parcels of land to build entire communities in assembly-line fashion. A postwar example noted in Chap. 1 was William Levitt's Levittown community of almost 18,000 houses, built on Long Island, NY. Before World War II, 86 % of all American homebuilders built less than three houses per year. By 1959 large volume homebuilders such as Levitt accounted for the creation of two-thirds of all new single-family houses.[5]

Fig. 2.7 By 1959, large volume homebuilders were responsible for two-thirds of all new single houses. During prewar times, homebuilders would be completing less than three houses per year

Within this context of mega-builders, nonfinancial concerns, such as livability, comfort, and human scale, may not be addressed as well as with a folksy neighborhood homebuilder. In the big business world of homebuilding, houses are designed merely as "products," and the marketing and sale of products is the objective of most any business; thus, it may be more difficult to assess and prioritize "community design" and other nebulous objectives, unless they can be shown to increase the financial goals of the corporation.

While large, high volume homebuilders continue to dominate the industry, many of them regard quality design as part of their corporate policy. Historically, community developers such as J. C. Nichols, who developed the Country Club District discussed in Chap. 1, show that large-scale builders can incorporate quality design into their business model. Today, large national homebuilders such as Texas-based David Weekley Homes address each region's market with small lot designs that include front porches and other human scale detailing local to the region. Charter Homes of Lancaster, PA builds communities that preserve farmland and the region's natural ecology. John Wieland Homes in Atlanta, GA builds a percentage of the company's houses at below-market prices to help address the need for affordable housing.

Change Comes Slowly in the Housing Industry

One of the biggest challenges in bringing about improved housing design standards is overcoming conservative attitudes of builders, public officials, and citizens. Builders generally don't want to try something that hasn't already been built enough times to guarantee proven marketability. Since they put financial resources at risk each time a new home or community is undertaken, it is understandable that conservative attitudes tend to prevail.

In spite of the high financial risks associated with change innovation does occur and calculated, well-researched, progressive design is reasonable and prudent to pursue. Typically, we find that medium-sized builders—those who have more financial leeway than smaller builders, but less corporate dogma than larger builders—undertake these new ideas. I have also noted that in some geographic markets, builders are more willing to experiment with new design ideas than in others. Once an innovative project is proven successful, however, there is often an incentive to emulate the concept by other homebuilders, large and small.

Fig. 2.8 (**a**) Texas-based David Weekley Homes, which builds in communities nationwide, tailors' designs for the regional vernacular, as seen in this Craftsman style house in North Carolina. (**b**) Charter Homes, a Lancaster, PA-based regional builder, focuses on developing unique mixed-use and walkable communities

Do Buyers Really Have Options?

Another major problematic attitude within the housing industry is the dogma that *if a product sells, then it is a good product*. This is an interesting topic for philosophical discussion: do people buy houses with poor or mediocre design because they like them, or because it's all they can get?

To me, the latter is more likely to be true—*buyers need to take what they can get* in the housing market, particularly shoppers that need modest-priced houses. People do not prefer to buy ugly houses, but they do need or want to buy something. Consider some of the typical issues affecting the decision of the average buyer of a new house:

1. *Price*. This is the number one factor limiting buyer choice. Many households are on a tight budget, and a poor exterior appearance of a house may be a necessary trade-off for ownership.

2. *Familiarity*. When making one of their largest investment ever, buyers tend to be conservative. Older plans that are prolific in the market and familiar tend to be strong sellers, as some buyers like plans similar to ones they grew up in.

3. *Comparables*. When almost all new homes in a market are of poor design, buyers start to feel that they have few options. And when there isn't a well-designed community in the area as an example, everyone accepts the lower standard as a given.

4. *Quality Construction*. A house can be very well constructed and still be poorly designed. Buyers generally recognize solid, quality construction, but less easily discern quality design, so a house with high-quality materials and specifications may be purchased in spite of poor community or interior design standards.

Evidently, there are plenty of reasons why suburban housing has been the focus of harsh criticism from many viewpoints. Let's now move on to some positive responses to these issues.

Traditional Exteriors/Contemporary Interiors

As noted, consumers may need to settle for whatever designs are available, particularly when it comes to the affordable category. But when questioned in focus groups, potential homebuyers show considerable yearning for higher quality designs. Several years ago, our firm was commissioned to design a series of models for a new community developed by Davidson College, in Davidson, North Carolina, in order to assist with housing its faculty and administration. As in common practice, a consultant hired to program the designs held focus groups of potential buyers to receive their input on how the new houses should look and be designed.

Fig. 2.9 Model houses designed for *McConnell at Davidson*, a new community sponsored by Davidson College in Davidson, NC, emulate turn-of-the-century styles found in historic Davidson. By polling potential buyers, the developer was made aware of a preference for historical styles over typical tract houses

What cames out of the focus groups was very interesting. People clearly indicated that they would like to be able to live in homes that were more like the turn-of-the-century houses in the town and specifically would not want to buy models typically found in new communities. In response, a series of designs emulating historical styles was developed and built for the homebuyers.

The New Urbanism: A Reform Movement

As a result of the growing dissatisfaction with postwar community planning, new design philosophies along the lines of the Davidson focus group scenario began to emerge during the 1980s. One approach suggested modeling new communities after historical small towns, called *traditional neighborhood developments* or TNDs. This view was advocated by several design practitioners—most notably by Duany/Plater-Zyberk Architects in Miami, Florida, Looney Ricks Kiss Architects in Memphis, Tennessee and Calthorpe Associates in San Francisco, California. Since then, many additional design firms throughout the country have adopted this design philosophy and planning officials have come to embrace these concepts as well.

These planners suggested that new community designs should emulate mixed-use, pedestrian-oriented environments similar to Annapolis, Maryland or Carmel, California. Andres Duany and Elizabeth Plater-Zyberk got ahead of the game because their plan for developer Robert Davis was largely built first. Seaside, the 80-acre resort hamlet in Florida's panhandle, made quite an impression with the architectural and building press, as well as with the popular media. For a population that was dissatisfied with the suburbs, this was a breakthrough development hailed as the antidote to sprawl.

Fig. 2.10 *Seaside*, the 80-acre resort hamlet on Florida's panhandle, made quite a media splash as a traditionally planned community. Note below the circulation concept extending to the beach and ocean

Fig. 2.11 DPZ's plan for *Seaside* was inspired by the mixed-use, pedestrian-oriented coastal towns of the Southeastern US. The design formula included civic architecture and meaningful open spaces in key locations, traditional and sustainable building techniques, a regulating plan, and an urban design code. (Courtesy Duany Plater-Zyberk & Company)

Seaside, conceived in the early 1980s, heavily influenced production housing. Davis' commitment to actually build the community as a departure from the status quo added credibility to the TND model. The design formula used at *Seaside* emulates the charm of turn-of-the-century villages with an Olmsted-like master plan that includes civic architecture, open space, and an urban design code which allowed for a mixture of both highly nostalgic and avant-garde architecture to coexist. Criticism of *Seaside* focused on its elitist character as a second home community, and labeled traditionalist planning as another form of social engineering. However, Seaside is still recognized for its early role in identifying the need for new planning models for suburban communities.

Another early traditionalist community that addressed more affordable housing is *Harbor Town*, built on an island in the Mississippi River in Memphis, Tennessee. *Harbor Town* was impressive because it comprised mainstream primary housing for varied household income groups, as opposed to *Seaside*'s second-home/resort tenure. *Harbor Town* was master planned by RTKL Architects of Baltimore, Maryland, with the initial architecture provided by Looney Ricks Kiss Architects of Memphis. The neighborhoods at *Harbor Town* embody traditional community design patterns with homes organized along sidewalks and open greens, while cars and parking are relegated to rear alleys. Based on the success of *Harbor Town*, Looney Ricks Kiss went on to plan and design many traditionalist communities throughout the USA.

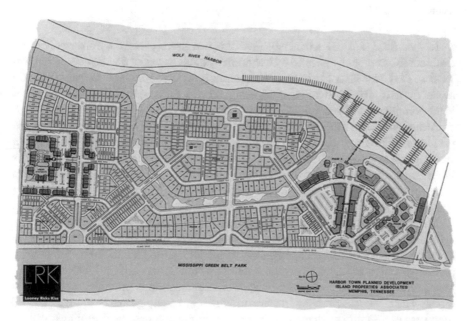

Fig. 2.12 *Harbor Town* in Memphis, Tennessee is a traditionalist community that addresses more affordable primary housing designed by Looney Ricks Kiss Architects. The original master plan by RTKL Associates, with modifications and subsequent phases by Looney Ricks Kiss Architects, stressed pedestrian circulation as well as mixed-use and mixed-income housing types. (Courtesy Looney Ricks Kiss)

Fig. 2.13 These houses at *Harbor Town* include generous porches in the front and garages in the rear of the lot creating an appealing street scene. (Courtesy Looney Ricks Kiss. Photo by Jeffrey Jacobs, Mims Studio)

Laguna West, outside Sacramento, California was an early west coast primary housing traditionalist community planned by Calthorpe and Associates of San Francisco. *Laguna West* was to be the first of visionary Peter Calthorpe's *transit-oriented designs* (TODs), a model for new mixed-use communities designed such that residents can walk anywhere in the community in 15 min and also have mass-transit access to the greater region.

Although many aspects of the original plan were not realized, including the mass-transit link and the office/commercial buildings near the entrance, *Laguna West* raised awareness of the need to experiment with alternative settlement patterns including mixed-use, walkable communities with transit access. Calthorpe's firm went on to plan many successful communities based on these concepts including *Stapleton* in Denver, CO and *Northwest Landing* outside of Seattle, WA.

These early TND and TOD communities proved to be highly influential and were followed by a plethora of similar designs nationwide. Andres Duany and other practitioners later published comprehensive rules and principles for new community planning, which eventually became known as New Urbanism. In 1993, this small group of design professionals established an organization to promulgate these principles, The Congress for New Urbanism, which has since grown to close to 20,000 members representing a diverse collection of public and private disciplines.

The Congress for New Urbanism (CNU) and its members has made a tremendous positive impact on new community design here in the USA, as well as worldwide. The publication of regulating tools, such as the Smart Code and Formed Based Zoning, both of which dictate the size and shape of buildings as opposed to use, has been adopted by a number of local governments as a means of improving new

Fig. 2.14 In California, the *Laguna West* community in Elk Grove outside Sacramento is a West Coast traditionalist community, master planned by Calthorpe Associates of San Francisco. The plan includes mixed-use, varied housing types, and a community center near a proposed mass transit connection. (Aerial imagery © Google Maps)

community and neighborhood design. Through annual conferences, workshop, and seminars, the CNU has become a highly influential force in the design community.

Over time, other national design and planning organizations such as the Urban Land Institute, the American Planning Association, the U.S. Green Building Council, and the American Institute of Architects recognized the general principles of the New Urbanism as appropriate directions for sustainable new community design. These principles advocate for more equitable, socially integrated, and overall more sustainable communities.

Sustainable Communities

Peter Calthorpe has also been a prolific author on the movement toward sustainable planning and design. In his 1986 book, <u>Sustainable Communities</u>, coauthored with former California state architect Sim Van der Ryn and other contributors, Calthorpe provided guidelines for designing settlements that are less dependent on automobiles and nonrenewable resources.[6] This early book, republished in 2008, argued for increased consideration of environmental issues in new housing and community designs—including house orientations for solar power, using local renewable materials for construction and planning for individual on-site food gardens.

Current emphasis on sustainable design is still focused on using renewable resources for construction, alternative energy systems such as solar and geothermal, low-water use fixtures and landscaping and "smart house" appliance controls to reduce energy. The principles outlined in <u>Sustainable Communities</u> are consistent and compatible with social community design objectives: mixed-use, moderate density, self-sufficient settlements linked to larger cities with mass-transit, built with

Fig. 2.15 Many of Peter Calthorpe's other TODs, or transit-oriented developments, followed in the planning footsteps of *Laguna West* to create vibrant and connected communities such as at *Northwest Landing* in Dupont, WA. (Courtesy Calthorpe Associates)

renewable resources and designed to reduce the use of polluting and nonrenewable resources. We will discuss these issues in more depth in Chap. 10 of this book.

Other Recent Planning Movements

In addition to New Urbanism, other improved community design practices have evolved to address the negatives of suburban sprawl. Most of these are cataloged in planner Randall Arendt's <u>Rural by Design: Planning for Town and Country</u>. Based on years of practice and travel in working with communities throughout the country, the book chronicles design concepts ranging from low-density conservation and farm communities to downtown and mixed-use settings.[7]

Most of the low- to medium-density communities in the book address building houses on only one section of a parcel, leaving the balance in common recreational areas such as parks and gardens. This is similar to golf course communities, which were perhaps the most popular theme for new housing development targeting afflu-ent buyers over the past several decades. At this point, though, the demographics for golf are fading, while others are gaining traction. One of these themes is the farm-based community, where new homes are located on a portion of the land, allowing the bulk of the property to remain in an agricultural state. In some cases, residents can share in the produce grown on the farm. This idea captures the still-popular "buy fresh, buy local" movement.

Fig. 2.16 *Willowsford,* a new community outside of Washington, D.C, spans 4000 acres, with half of the land conserved in a natural state or preserved as working farms, with local produce available to residents

Other conservation communities leave a significant portion of the site in a natural state, such as a forest or wetlands. The infrastructure for the housing may have a less urban character, with some streets left without sidewalks or curbs in favor of trails though the open spaces. The density and relationship of the houses to the street and to each other would generally follow the general principles of community design found in a typical well-designed community such as those discussed under New Urbanism.

Still other significant planning movements that have evolved over the past several decades include Smart Growth, Transit-Oriented Development, and Walkable Communities. All of these have similar goals to the New Urbanism and Sustainable Communities, in that they call for moderate- to higher density housing that will support services within walking distance, thereby decrease reliance on the automobile, or allowing walkable or bike-able access to public transit. Smart Growth in particular favors development in areas with existing public services such as roads and utilities already installed, called "brownfield" areas, as opposed to expanding onto the periphery and extending into "greenfield" locations. These movements also advocate for housing designs that conserve natural resource consumption and utilize renewable energy sources.

Traditionalist and Classicist Architects

Traditionally planned communities do not necessarily depend on traditional architecture. In historical suburbs, a period architectural theme was often incorporated, as with Forest Hills Gardens with its English country theme. Early houses at *Seaside* were generally in the traditional Florida "cracker" style though later buildings departed from traditional designs.

Fig. 2.17 The popularity of traditional architecture is reinforced by the prevalence of design firms with traditional designs as their calling card. One such instance, Stephen Fuller's Atlanta-based firm, has clients nationwide. (Courtesy Stephen Fuller Designs)

Furthermore, traditional planning does not depend on traditional architecture to be successful. Many architectural themes associated with good community design need not be stylistically confined. Style is of no issue to concepts such as recessed garages, porches, courtyards, and other architectural elements that connect housing prototypes to the greater community.

I must point out, however, that the popularity of traditional architecture in the residential market is virtually constant. Currently, many of the largest stock house plan services and design firms throughout the country use traditional architecture as their calling card. Browsing through house and garden magazines one finds plan services in firms named Historical Concepts, Design Traditions (now Stephen Fuller Designs), and Southern Living Homes. The designs these organizations produce fill a much-sought-after look that homebuyers respond to, confirming the public's overwhelming interest in traditional residential architecture.

In addition to the consumer interest in traditional architecture, the art of classical design has also gained momentum in recent decades. In response to the heavy favoring of modernism by the design community and media, a few schools of architecture began returning to teaching classical architecture principles dating back to the Greek and Roman eras and the publications of Vitruvius. In 1968, the Classical America organization was founded; in 1991 the Institute for Classical Architecture was established. The two organizations merged in 2002 and became the Institute of Classical Architecture and Art, or ICAA.

Fig. 2.18 Traditional New England styles were used at *The Meadows at Riverbend*, a new home community in Middletown, CT. The land plan also has a state-of-the-art stormwater management system

The ICAA, based in New York, is a teaching and advocacy organization with local chapters nationwide. The membership includes some of the leading voices in the CNU, such as Andres Duany, Elizabeth Plater-Zyberk, Robert A.M. Stern, and Leon Krier. The ICAA, along with the CNU and the schools of architecture that continue to include the study of classical architecture, has acted as a counterbalance to the modernism preferred by mainstream architects.

One might note that achieving the exact proportions and quality of materials preferred by classical architects may be out of reach in terms of cost for basic production housing designs. However, if classical design is considered a subset of traditional, historical, or vernacular residential architecture, it is helpful to understand classical design principles and follow them to the extent that they are justifiable in the current housing market.

As a result of these influences, we see more new home communities adopting a particular historical flavor, such as the bungalow-style houses designed for *McConnell* in Davidson, North Carolina, or the prototypes for *Meadows at Riverbend* in Middletown, Connecticut, where designs are borrowed from traditional New England Saltboxes, Garrisons, and Cape Cods. And homebuyers continue to respond to such traditional residential architecture since it is recognizable and familiar.

Humanistic and Interior Design Principles

Other recommended solutions to improve current planning practices have been offered by architect/scientist Christopher Alexander and his colleagues at the Center for Environmental Structure at the University of California at Berkeley. Alexander advocates a system of planning and design implemented by master builders who empower people to build with time-honored "patterns" based on human and social needs. These patterns are cataloged in a bible-like book called <u>A Pattern Language</u>, which, along with other books written by Alexander and associates, describe a more universal concept of dwelling, neighborhood, community, and city.

In addition to planning principles, Alexander's A Pattern Language provides guidance for interiors layouts and details to be applied to the design of the houses within the community. Suggestions for alcoves, built-in seating, etc. are explained in detail in mini-chapters of the book.

In addition to A Pattern Language, there is no shortage of literature on making housing designs more livable, enjoyable, and charming. Some of Christopher Alexander's colleagues have published their own books along these lines, as have a myriad of other residential architects, including notable practitioners such as Charles Moore and Donlyn Lyndon. One influential book by architect Sarah Susanka, The Not-So-Big House, snowballed into a series of books extolling the virtues of living in smaller, but better-designed homes. Many of these titles are listed in the bibliography of this book. However, most provide ideas for the more expensive custom home market instead of the cost-conscious production housing.

Production Housing Priorities Moving Forward

Unlike the postwar building boom scenario, residential builders are now responding to issues beyond providing basic shelter. Production housing is shifting its focus to address reduced energy consumption, sustainability issues, and improved community design, as well as the need for well-designed affordable housing—for which there is an almost unlimited demand, provided it can be built in a comfortable, livable setting.

Fig. 2.19 Ryan Homes, one of the largest homebuilders nationwide, features a *BuiltSmart* option, which offers increased conservation of natural resources and energy, comfort, quality, and environmental protection in the building process of new homes and communities

America's housing delivery system is the envy of the free world: statistically and technically, we are the best-housed nation on earth, and yet there is more to be done. The lower economic strata of our society remain ill housed, particularly in comparison to other industrial nations. Our new communities are still economically and racially segregated and wasteful in their energy consumption.

As we move forward public officials, planners, design professionals, and builders of new homes and communities will all work toward addressing these challenges. The goals will be to produce neighborhoods that are more diverse and less dependent on the automobile, with houses that will be more livable with enjoyable spaces in which to reside.

Notes

1. Siniavskaia, Natalia. *Eye on Housing*. "14 Million Households 'Priced Out' by Government Regulation" National Association of Home Builders: 12 May 2016.
2. Putnam, Robert D. Bowling Alone: The Collapse and Revival of American Community. Simon & Schuster: New York, NY, 2000.
3. Watson, Ray. 1989. Presentation to AIA Housing Committee in Newport Beach, CA.
4. "Housing's Contribution to Gross Domestic Product (GDP)". National Association of Home Builders. Washington, DC.
5. Weiss, Marc A. "The Rise of the Community Builders: The American Real Estate Industry and Urban Land Planning". New York: Columbia University Press: 1987.
6. Van der Ryn, Sim and Peter Calthorpe. Sustainable Communities. Sierra Club Books: San Francisco, CA, 1986.
7. Arendt, Randall. Rural By Design: Planning for Town and Country. American Planning Association Planners Press: Washington, DC, 2015.

Chapter 3
Community Planning and Design

Problem solving generally starts with the broad brush issues. In order to address how we can make our new housing more livable, and our neighborhoods more attractive, let's start with a discussion of new community planning.

Community planners and land planners come from a variety of backgrounds, which has been subject to change over the years. While one may initially think most residential communities were laid out by quintessential planners such as Frederick Law Olmsted or John Nolen, unfortunately that was not the case. In reality, the average subdivision did not warrant the cost of that level of design expertise.

Land planners typically come from one of three backgrounds—civil engineering, landscape architecture, or architecture. Planners with degrees in Urban and Regional Planning, Urban Design, or other planning programs have also entered the field; however, these professionals are generally educated in curriculums that are subsets of one of the above disciplines.

As a result of these differences in training, planners tend to have divergent priorities in designing new communities. The civil engineers tend to focus on practical issues such as storm water drainage and sewer pipes—instead of orienting homes for livability concerns. Landscape architects, on the other hand, tend to emphasize new and natural planting areas over the design of the houses. Architects, in turn, sometimes ignore the above issues and focus on the design of the houses. As a result of these divergent priorities, new communities can often look like they were designed by committee. And they were--many new communities tend to be hybrids of different views that should, but don't, always mesh together.

However, typically community design is most influenced by civil engineers. When community developers buy a parcel of land, they know that obtaining the crucial engineering and zoning approvals is a highly political process. Therefore,

Fig. 3.1 How can new communities achieve qualities such as the reasonable street widths and lot sizes seen in this prewar neighborhood in Oklahoma City?

© Springer International Publishing AG 2017
J. Wentling, *Designing a Place Called Home*,
DOI 10.1007/978-3-319-47917-0_3

they need to hire a civil engineer familiar with local controls over land development, typically a politically connected engineering firm.

In securing planning and zoning approvals, the key issues are generally focused on mitigating the impact of new development on traffic and local government services, such as trash collection, fire safety, and school enrollment. Other design considerations—such as whether the proposed plan has moderate size streets and usable open space—generally are not as important.

Once a planning submittal reaches a certain size, proposed urban or community design schemes may be reviewed in the application process. But these are generally considered lightweight issues, certainly not significant enough to delay or jeopardize approvals. Some local governments require that even small subdivisions include architectural review, but again this is typically a cursory procedure. What we need in the approval process is a more holistic and balanced view of which community planning issues should be subject to review, and the professionals who can lead the development team and respond to those concerns.

Firms mentioned earlier such as Duany/Plater-Zyberk of Miami, Calthorpe and Associates in San Francisco, and Memphis-based Looney Ricks Kiss, among others, have been practicing an integrated approach to community planning that balances the concerns of architecture, engineering, and landscape design in a fashion similar to Olmsted and Vaux. Although these firms are taking different approaches in their work, they and most progressive firms are advocating the same general principles of community design that include the following:

1. *Mixed-income and mixed-density housing*
2. *Pedestrian-oriented streets and circulation systems*
3. *Higher density housing with quality open space*
4. *Mass-transit links to larger cities*
5. *Integration of residential and commercial uses to reduce dependency on the automobile*
6. *Environmentally responsible planning and design*

In many cases, these planning principles are described in journals as *neo-traditionalism*; however, the goals stated above should not be viewed as a trend, but rather as a return to acknowledged community planning goals that have been practiced for centuries, and as recently as the 1950s.

Fig. 3.2 (**a**) The New England village with its mixed uses and diverse architecture is in stark contrast to the segmented, homogeneous planned unit developments of today (**b**)

Integrated Village Vs. Segmented Planned Unit Developments

Most post-World War II communities were shaped by the early zoning dogma of *separation of uses*—which called for the complete segregation of residential land from commercial and/or industrial uses (validated by the courts in 1926 through the landmark case, *Village of Euclid* vs. *Ambler Realty Co.*)[1] Initially justified by the perceived need to protect residential neighborhoods from encroaching factories moving out of city cores, this doctrine was carried to absurd conclusions—which is why in most of our suburban communities it is impossible to buy a loaf of bread without getting into a car and driving for 15 min to get to a store.

The theory behind separation of uses eventually led the residential design community to introduce the concept of the "planned unit development" (PUD) in the early 1960s. This terminology later evolved into "master-planned community" (MPC). Originally conceived to offer developers more design flexibility than rigid zoning standards would, the MPC would become a stock recipe for community design. Too often an MPC would be conceptualized as follows: spread a piece of paper over a site survey. Start drawing circles. Label those circles with different housing types—large homes, small homes, townhomes, apartments. Now connect the circles with lines to the access road to form the circulation system. There you have it—a segmented MPC designed for people who want to live in segmented environments.

Perhaps as a result of the extreme to which segmented MPCs have been taken, current homebuyer surveys and focus groups now indicate a preference for homes that are located in neighborhoods with mixed uses, different housing types, and varied economic groups. Potential homebuyers want to live in walkable neighborhoods with sidewalks and access to transit to reduce the need for a car.[2] Therefore, the current strategy for building successful new communities would favor creating new homes and neighborhoods that capture the human-scale spirit of the older ones.

This is not to say that new communities need to be Disneyland-like replications of turn-of-the-century villages (which is too often the case). Instead, we should look to the underlying principles of planning and design found in these small town examples that are still applicable to the human and social needs of contemporary society including mixed housing types, integrated commercial services, pedestrian circulation, open space design, and public transportation links to metropolitan centers.

Mix of Housing Types

One of the tenets of MPC design is to separate different housing types into different "pods" or "land bays" defined by the master plan. Townhouses over here, single homes over there, and so on. This works very well for development administration and construction sequencing since most participants in the homebuilding industry are specialists. From the marketing point of view, there are often arguments in favor of not mixing housing types.

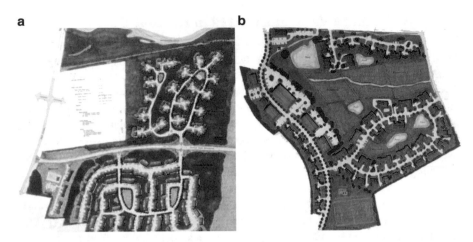

Fig. 3.3 (**a**) Communities that are built with primarily one building type can be monotonous, while a mixture of building types, land uses, and architectural styles can provide more vitality for communities (**b**)

However, the goal of segmentation that drives MPC design contrasts with the more integrated uses and architectural variety found in traditional towns and prewar suburban plans. Here, one often finds a mixture of housing types—single homes, duplexes, townhouses, and garden buildings—within blocks of one another or on the same street. This mixture of housing stock gives a neighborhood and community an architectural diversity and social vitality by providing housing opportunities for a variety of household compositions and income groups.

Looking at a street that may switch from single homes to duplexes, or triplexes, we see a rhythm of shapes and masses that breaks monolithic pattern. From a marketing viewpoint, offering a varied program of sizes and shapes in houses helps to broaden the target market of the community; some people want smaller attached homes, while others prefer detached. Some people want the single-level living offered in garden buildings, while others want the vertical floor plans offered in townhomes.

Homebuilders typically have avoided mixing housing types within a new community due to the fear of downgrading the value of more expensive houses by placing them next to less expensive ones. While this is a legitimate concern, design controls and coordinated planning and architecture can help achieve a mixed-density community that looks and feels appropriate. It is true, for example, that you may not want to locate the highest priced single-family model right next to the lowest price garden unit. However, it shouldn't be a problem to locate a single home next to a comparably priced townhome or garden unit, or next to a smaller single home with 30 % less square footage.

One example of mixing housing types at a new community is *Summerfield at Elverson*, which lies on the fringe of a small borough of 750 people in Southeastern Pennsylvania. In order to develop a 200-acre farm that would roughly double the size of the current borough, we helped to develop a master plan that included a broad range of housing types similar to the mix of housing found in historical

Fig. 3.4 (**a**) *Summerfield*, a community outside Elverson, PA, mixes detached houses with attached townhouses on the same street to continue the varied housing pattern of historical Elverson. (**b**) Housing prototypes at *Summerfield* are based on German Colonial and Carpenter Gothic styles found in Elverson

Elverson. The planning approach combined single houses with duplexes and townhouses, sited together along streets within each neighborhood.

The architecture at *Summerfield* was crafted based on historical styles found in the village. Carpenter Gothic and German Colonial homes were studied and interpreted into contemporary designs, which were then carefully integrated to create a diverse community similar to the original town.

Community developers that are restricted to building single houses by way of zoning or other controls may want to consider varying their lot sizes to introduce architectural variety into the neighborhood. By mixing lot widths along a street, developers can encourage houses of different sizes to be built adjacent to one another. Lots with the same depth, for example, may be sized with widths that vary from 60 to 90 ft, allowing houses with a different character to formulate the streetscape. Houses on lots with narrow footprints may have plans locating the garage in the rear of the lot. Houses on wide lots, on the other hand, may include side-load garages.

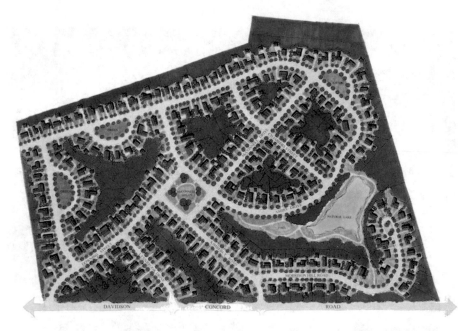

Fig. 3.5 At the *McConnell at Davidson* community, lot sizes were varied to help create diversity along the streets. Lot widths ranged from 65 to 90 ft to promote the construction of different house designs. The community also features open greens, playgrounds, a natural lake, and nature trails

Mixing housing types and lot sizes within a community, neighborhood, or street helps add vitality to a residential setting. It will allow for not only architectural diversity, but also for the interaction of somewhat different income and age groups. Although some residential communities have seen success in restricting their market to a single group (such as retirement communities), the overwhelming precedent of most historical residential communities has been diversity.

Neighborhood Commercial

One of the tenets of traditionalist planning is integrating retail and services into the community plan based on older neighborhood models of local, convenient retail services. On every other corner of urban neighborhoods, there is a small grocery store, drugstore, dry cleaner, or other service business—that are typically family-owned enterprises—that serves the local neighborhood's needs.

With the introduction of zoning and the doctrine of separation of uses, corner stores were outlawed. Absurd parking requirements for commercial uses were established. Meanwhile, the retail industry evolved toward stocking larger facilities with lower priced goods based on the economies of scale permitted by automobiles. Therefore, the primary retail prototype of the postwar era became the strip shopping center, usually located off a major roadway that was accessible only by car.

Fig. 3.6 (a) Neighborhood commercial such as this corner store were absent in early suburban communities. (b) Here, The Market—in the mixed-use Village Green of the master-planned community The Pinehills in Plymouth, MA—provides convenient groceries, fresh produce, meat, fish, and prepared foods in 14,000 SF serving Pinehills residents and the surrounding community. (Photo (b) courtesy Pinehills, LLC)

New community master plans often address the need for neighborhood retail by including a commercial parcel along the road frontage portion of the property. Even with this approach, though, shopping typically still requires a car due to the lack of pedestrian connection between the residential and commercial parcels. What we need is a still finer scale of retail facility that can be integrated directly into the residential fabric of a neighborhood.

The postwar iteration of the corner store is the ubiquitous stop-and-shop type facility found at roadway intersections or gas stations. If these businesses could develop prototypes to fit on a typical residential corner lot, neighborhood retail could be readily woven into community master-planning. Designs would have to mitigate issues such as parking, traffic, and noise—and planning boards would have to reduce typical commercial zoning requirements (primarily for parking spaces)—but the resulting design would be a beneficial "walk-to-the-store" amenity for new communities.

Some community developers have begun to sponsor general store-like convenience shops within communities, an example of which can be found at the center of *Southern Village* in Chapel Hill, NC, a community by Bryan Properties.

Developers are working with large grocery chains to sponsor smaller prototype shops that can fit into their neighborhood fabric. If these can be viewed financially in the aggregate with their larger, more economically scaled profit centers, walk-to-shop facilities will be returning to our neighborhoods. The integration of retail and residential, although challenging, definitely is possible and is becoming more common in future residential communities.

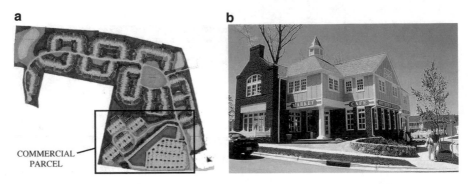

Fig. 3.7 (**a**) Community master plans generally group neighborhood retail at the front of the property and accessible only by car. (**b**) Logistically, commercial space can be accommodated on typical corner lots when properly designed, as seen in Bryan Properties' corner market inside *Southern Village* in Chapel Hill, NC

Narrow Vs. Wide Streets

Readily apparent in prewar residential subdivisions are the more reasonably sized, pedestrian-oriented streets; older neighborhood streets may be around 20 ft wide, compared to new streets, which typically can be up to 36 ft or wider. The obvious reason for the smaller width of the older streets is they were built before the current emphasis on the automobile. Many medium density prewar communities also commonly have sidewalks on both sides of the street, often with mature shade trees.

While new communities tend to have wider roads, sidewalks and street trees are often omitted. One reason for this is that many older streets were built and paid for by the municipal government as an investment to encourage housing to be built, which in turn would increase the municipal tax base. In today's no-growth atmosphere, the private developer is forced to pay for street construction, which the local government has now mandated to be much wider. So, to compensate for the increased cost of building the wider streets, developers lobbied to omit the sidewalks and street trees.

Excessively wide streets destroy the sense of human scale in residential communities. They also waste valuable land, increase impervious surfaces and storm water runoff, encourage speeding cars, and discourage walking. In order to address this problem, many planners and community developers are (1) working with municipalities to revise public street design standards and (2) implementing more private street systems.

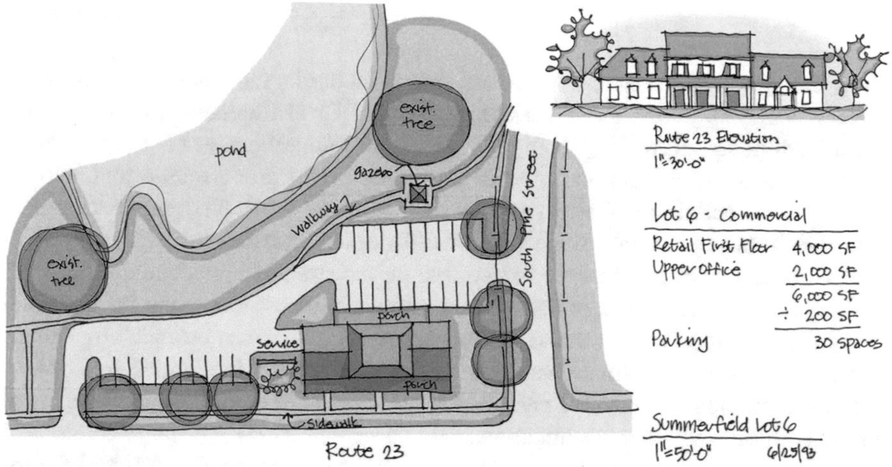
pond
exist. tree
exist. tree
gazebo
walkway
South Pine Street
porch
service
porch
sidewalk
Route 23
Route 23 Elevation
1"=30'-0"
Lot 6 - Commercial
Retail First Floor 4,000 SF
Upper Office 2,000 SF
6,000 SF
÷ 200 SF
Parking 30 spaces
Summerfield Lot 6
1"=50'-0" 6/25/93

Fig. 3.8 Neighborhood commercial at *Summerfield* has road frontage visibility as well as pedestrian access from residential sections. The residential scale of the building is in character with the nearby houses

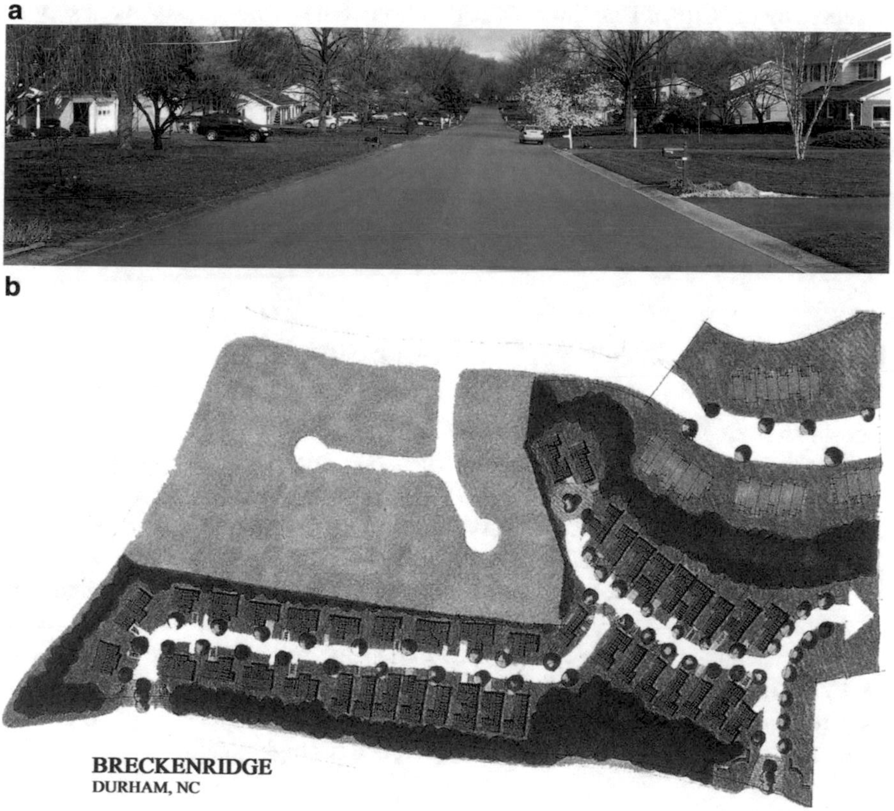
a
b
BRECKENRIDGE
DURHAM, NC

Fig. 3.9 (**a**) Excessively wide streets destroy human scale in residential neighborhoods, increase impervious surface coverage on site, encourage speeding cars, and waste space. (**b**) At a neighborhood in Durham, NC, narrower 24-foot-wide private streets were approved to reduce development costs

In order to promote more reasonable street standards, the National Association of Homebuilders joined the American Society of Civil Engineers and the Urban Land Institute to publish <u>Residential Streets</u>. This comprehensive book advocates a series of roadway designs matched to the amount of traffic volume that will be served—with certain local street widths as narrow as 22 ft. The publication also addresses more reasonable standards for cul-de-sacs, turnarounds, curbing, sidewalks, and a host of other issues to improve public streets in residential communities.[3] If public street standards cannot be altered, private streets can be built and maintained by the developer and its successor, which means establishing homeowner association to maintain and insure the roadway after the homes are sold. In exchange, municipalities will sometimes approve less stringent designs in terms of width, curbing, and other specifications.

The benefits of private streets can be seen at *Breckenridge*, an affordable community in Durham, North Carolina. In order to reduce both construction costs and additional land consumption, *Breckenridge* was built with private roads that were only 24 ft wide with an inverted pitch (surface water drains to the center of the street) to avoid the need for curbs and gutters. These reduced design standards were accepted by the City of Durham in consideration of the homebuilders' modest pricing of the houses.

Private streets are owned and maintained by homeowner associations. Typically, the smaller the development, the more difficult it is to market a community with private streets—due to the single-family buyer resistance to homeowner associations. Therefore, a true resolution to the problem of reducing street widths lies with local government adoption of more reasonable, human-scaled dimensions for public streets.

Interconnected Streets Vs. Cul-de-sac Streets

One of the big items of debate between traditionalist and mainstream planners focuses on the benefits of dead-end streets vs. connected or through streets. Conventional wisdom states that people like to live on a quiet, private street with little or no cross-traffic; hence, ideally all houses should be located on dead-end cul-de-sacs accessed from collector roads.

The traditionalists criticize dead-end streets as a circulation system for a variety of reasons. First, cul-de-sacs tend to form isolated pockets of houses connected to the larger neighborhood via a high-speed collector road. Thus, the negatives of heavy traffic may not be right in front of your home, but right down the street. A loop road system, on the other hand, consists of a web-like network of roads that connect to one another, none of which is designed for high speeds.

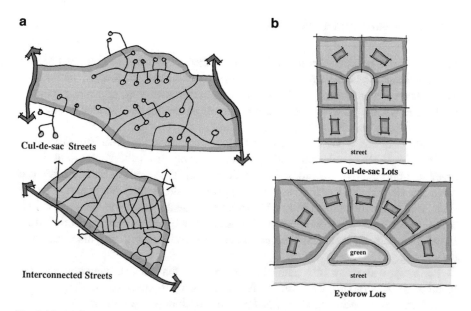

Fig. 3.10 (**a**) Cul-de-sac streets tend to form isolated pockets of houses, while interconnected streets weave together to form a more cohesive neighborhood. (**b**) Eyebrow streets can connect a row of houses to the greater community, while also providing a measure of privacy and territoriality

The second criticism that traditional planners have about cul-de-sac streets is that they do not work together to form a cohesive larger neighborhood. On cul-de-sac streets, a household is more likely to focus primarily on their immediate group of neighbors and have no reason to circulate to the next cul-de-sac. This creates small "territorials." Through streets, on the other hand, have continuity and connect to one another in a more comprehensive fabric, allowing more opportunities for social interaction.

Proponents of cul-de-sacs point out that in homebuyer surveys, families consistently favor houses on cul-de-sac streets because there is no through traffic—which enhances privacy and safety for young children. Perhaps a blended concept that draws off the benefits of the cul-de-sac and the through streets is the idea of eyebrow streets. Eyebrows can be public or private streets off a main street that serves a grouping of houses. Eyebrows connect the row of houses to the greater circulation system and community, but provide a measure of privacy and territory as well.

New Urbanist planners have studied further refinements to interconnected streets. Issues such as block length, points of intersection, and site lines can be addressed to promote walking, socialization, and visual quality. We have found in our practice, for example, that curved streets tend to enhance the streetscape view of a collection of houses better than straight-run grid street patterns.

While interconnected or through-road circulation patterns can and do have some negatives such as high-speed short-cutting or excessive through traffic, these problems can be mitigated with various buffers, setbacks, and other "traffic calming" devices which are found extensively in Europe.[4] Interconnected residential streets do make for a better circulation system in social terms of developing a sense of neighborhood within a community.

Vehicular and Pedestrian Circulation

Street design in residential communities would also benefit from a review of historical precedent for how to provide for both vehicular and pedestrian circulation.
Sidewalks were customarily built within the street right-of-way, adjacent or parallel
to the street curb. In many new communities, sidewalks are located around the
perimeter as jogging or walking paths or omitted altogether.

While I am not opposed to such perimeter paths, they play a different role from
the traditional street sidewalk and again move away from the social qualities of
older neighborhoods. The sidewalk location along the street is both a utilitarian
route around the neighborhood as well as a promenade. A sidewalk is part of a pub-
lic/private transition zone in front of the house where socialization is more likely to
occur. In communities without sidewalks, people will end up walking in the street.
Local government-mandated jogging trails, in comparison, are also generally
underused.

Perimeter walkways, bike paths, and other trails that connect new communities
to other neighborhoods, commercial facilities, or open space are desirable. They
should, however, be included in a pedestrian circulation system that addresses the
need for access, recreational and social needs, all scaled to the budget of the
community.

Fig. 3.11 The sidewalk location along the street is both a utilitarian route through a neighborhood
and a promenade, as seen at *Har-Ber Meadows*, outside Fayetteville, AR. By contrast, government-
mandated trails around the perimeter of sites are often underused. (Courtesy EDI International)

Small Vs. Large Lots

Typically, individual lot sizes in most eastern new communities are excessively large. In California and other Western states, lots are excessively small. What builders need is a happy medium between these extremes that can generally be seen in lots throughout the country dating from roughly 1900–1930, when the ratio between house size and lot seemed very appropriate and was a better example of balancing community and privacy concerns.

In the East and Midwest, lot sizes have grown since then, primarily for growth management, environmental, and other political reasons—but to the detriment of community design and affordability concerns. Typical lot sizes of a 1/4 acre or less comfortably accommodate modest-sized homes, yet some municipalities still have minimum lot sizes ranging from 1/2 to 2 acres.

Large lots do not help to define a neighborhood. When individual houses are too far apart they cease to relate to one another. While one-acre lots may be appropriate for upscale estate homes, they are really inappropriate for median price homes or affordable housing. Lots should be sized in proportion to the dimensions of the house, and a modest 2000 square foot house will easily fit on a 1/4-acre lot in rural settings, and a 7000-square foot (1/6-acre) lot in close-in suburban communities.

Fortunately, many municipalities have recognized that growth management objectives can be achieved by means other than large lots. An increasing number of community developers have responded to large-lot zoning by proposing *cluster housing ordinances*—under which the same yield of houses will be permitted only on smaller lots. This encourages the remaining lands to be preserved as common

Fig. 12 (**a**) Lots were too small in many Sunbelt markets, resulting in houses with narrow frontages dominated by garage doors. (**b**) These "wide/shallow" lot houses at Bressi Ranch built by Lennar outside of San Diego recess the garage and include front porches facing the street. (Photo (**b**) courtesy Tom Voss)

open space, which in turn can be creatively used to mitigate environmental concerns and to develop a unique sense of place or character for the community.

One example from our practice is *Old Farms*, an estate home community in Middletown, Connecticut. Here, the original 85-acre site was a horse farm zoned for 2-acre minimum lots. Graced with magnificent natural features including ponds, wooded lands, and panoramic views, the site also had extensive wetland areas that needed to remain undisturbed. Our planning proposal was to build the same houses that would be allowed with the two-acre zoning, but substitute one-acre lots and cluster the houses away from the sensitive wetlands and other natural features. The houses therefore could be organized around a large natural commons—with the front side of the houses facing the public commons and the rear of the houses facing the privacy of the woods.

Fig. 3.13 (**a**) Houses at *Old Farms* have views of open space used as pasturelands for horses. (**b**) A circa-1925 Sears, Roebuck & Company barn was restored for the boarding of horses, continuing the agricultural heritage of the land. (Photo (**a**) courtesy Robert C. Fusari, Sr)

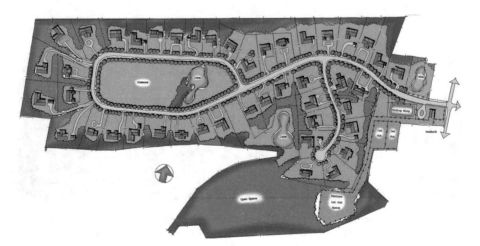

Fig. 3.14 At *Old Farms* in Middletown, CT, the two-acre minimum lot size mandated by zoning was reduced to one acre, in exchange for increased open space that allowed many homes to face a large commons

Sometimes, residual open space can also be used to carry forward the historical character of the land. In the case of *Old Farms*, the circa-1925 stable was restored and made available for residents to board horses. Open space such as pasturelands, riding trails, and a let-out area for the horses, all helped to maintain the original rural and agricultural character of the land.

While builders in the east and central part of the country are addressing large-lot zoning, California and parts of the west have the opposite problem: lots that have become the standard for new communities are too small. In response to the tremendous postwar pressure for moderately priced housing, Western local governments eventually began to approve new communities with lots as small as 3000 sq. ft. This decrease in lot size coincided with a corresponding *increase* in the average size of houses *and* the number of cars that were to be accommodated on the lot. For example, in 1950 a Western homebuilder might be putting up 1500 sq. ft homes with single car garages on 6000 sq. ft lots. Today, you see 2500 sq. ft homes with three-car garages on 3000 sq. ft lots. This translates to a floor-area-ratio increase of over 250 %!

Planners and architects in the 1980s tried to mitigate the impact of these density excesses with various house/lot combinations, including the "angled narrow lot," the "angled Z-lot," and the "wide-and-shallow" lot patterns. Most of these concepts did not really address the floor-area-ratio problem; however, privacy and community design were compromised in small-lot housing.

Perhaps the best small-lot housing prototype was the "wide-shallow" idea. These designs could be seen at the *California Meadows* in Fremont, California. Kaufman and Broad, a progressive California homebuilder, sited the houses that averaged 1600 sq. ft with two-car garages on lots as small as 3000 sq. ft. Although the lots are small, the lot/house designs incorporated 50-foot wide frontages, so that garages did not dominate the entry elevation. By using a concept for an interlocking rear lot line pattern with zero-lot-walls that the builder had copyrighted as "Zipper Lots," California Meadows was able to achieve a density of 8.7 units per acre and still maintain a quality street view for their houses. (See Figs. 4–14 in Chap. 4 for more on Zipper Lots)

Other planners are responding to the Western small-lot dilemma by reintroducing some of the lot-and-house patterns from earlier in the century—ideas such as rear garages, alleys, and shared driveways in an effort to develop more pleasant streets that are people and pedestrian oriented, particularly in some urban markets in Texas and the southwest.

In Chap. 4, I will talk more about small-lot/house patterns and in more detail and about the need to strike a balance in the need to reduce lot sizes in the East and increase them in the West (or at least organize them differently).

Common Vs. Private Open Space

One of the distinguishing characteristics of older subdivisions and communities was the frequent inclusion of open areas for common use. The best and most classical examples of this were the community greens, or commons, found in the villages throughout New England. Community planners as recently as the 1920s and 1930s would incorporate common space into residential subdivisions of various sizes as seen in places such as the Country Club District in Kansas City, discussed in Chap. 1.

Today, planners must use open space for purposes such as stormwater retention, buffer setbacks, wetlands, and other political/environmental purposes. Often the idea of devoting some land to socially usable open space is not given a high priority because the legally required open space has already taken up so much of the site that the developer must use all remaining buildable land for building lots in order to justify the project economically.

Furthermore, the inclusion of common open space generally requires a home-owner association to own and maintain the open space, sometimes an unpopular idea with buyers. This all-too-common argument against social open space is disappointing because even a very small common lawn can give a community a much-needed place for neighborhood gatherings. Most of the commons in New England,

Fig. 3.15 These modest houses in Tuxedo Park, NY face a common green that includes a playground

a   **b**

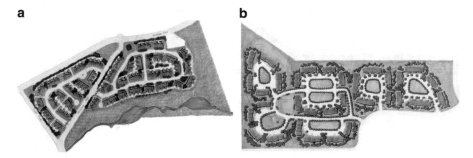

Fig. 3.16 (**a**) Most open space provided in new communities takes the form of buffers, wetland areas, stormwater retention, or other required greenery that are not socially usable by residents. (**b**) Some open space should be incorporated as greens or crescents that provide park views for at least some homes to give the overall neighborhood a sense of openness and relief

however, are maintained without homeowner associations or other legal entities. Planners visiting these small towns have noted that through various informal neighborhood cooperative efforts the lawn does get mowed. Homeowner associations were therefore rendered unnecessary.[5]

I cannot stress this enough: *a prime by-product of higher density housing must be higher quality open space*. We see this in our cities where, although people live at increased densities, they have at their disposal public parks, playgrounds, and other recreational amenities for common use. As new suburban planning boards adopt higher density ordinances, they must address the need for quality open space *within* the community, not at a remote location that requires driving.

Today, progressive developers and planners are incorporating pockets of open space into their designs. These are frequently reminiscent of the "park crescent" model inspired by architect John Nash's 1820s crescent of attached homes fronting on *Regents Park*. This simple example of a line of homes viewing a central lawn offers a wonderful model for contemporary communities. Not that every home in a community can be graced with a park view, but by introducing some common lawns into a land plan, the overall neighborhood will benefit from a sense of openness.

Open space should be distributed in new communities in a hierarchal fashion, which may include space for both active and passive uses, larger and smaller sized spaces, and natural and developed settings. A broad definition of open space should be encouraged that may include linear features such as sidewalks and bike trails, structures such as clubhouses or swimming pools, and nature preserves of undisturbed character. Programming the correct amount and type of open space is a critical planning decision that will vary with the location and needs of each community.

Architectural Controls Vs. Architectural Variation

Many new home communities built in the 1970s and 1980s experimented with architectural controls in varying degrees with the objective being to keep the architecture homogeneous. Examples include Irvine, California, The Woodlands, Texas and Reston, Virginia, all of which adopted contemporary design as the controlling motif. This approach contrasts to the architecture of traditional towns and villages that were built over a longer period of time by a wider variety of homebuilders in a myriad of different architectural styles.

While architectural controls can accomplish some positive objectives, they can just as easily result in a sterile environment. In general, the more architectural controls dictate a particular style, the more they result in mundane environments. Many of the modernist "new towns" outside London, Paris, and other European cities can illustrate controlled design dogma at its worst.

Design guidelines, a more liberal form of architectural controls, are frequently incorporated into larger and more upscale housing communities, such as golf course communities. Design guidelines can spell out which architectural styles are appropriate for a community, thereby often ruling out garish or eclectic architecture that

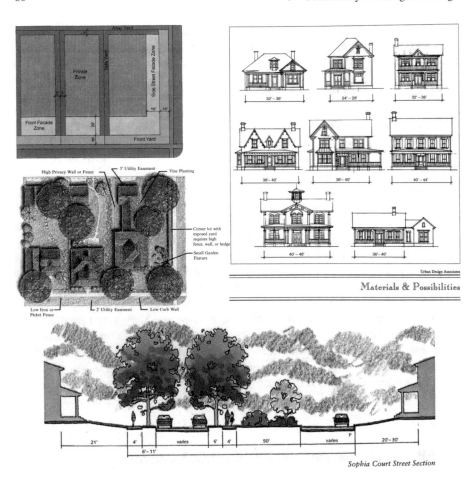

Fig. 3.17 The <u>Baxter Pattern Book</u> is highly visual and easily conveys community design objectives. Guidelines that suggest stylistic alternatives for houses of varied sizes are preferable to strict architectural controls that mandate one specific style or house type. (Courtesy Clear Springs Development)

may negatively affect property values. Guidelines that suggest stylistic alternatives are preferable to strict architectural controls that mandate a single style, but even design guidelines can still be oppressive if they are too narrowly drafted.

Some of the most successful design guidelines are those which are most simple and basic. Architectural guidelines for *Seaside*, written by Duany/Plater-Zyberk Architects, were straightforward and allowed for great variety in building expression. Urban Design Associates supplied homebuilder Clear Springs Development at the *Baxter* community in Fort Mill, South Carolina with a graphic of appropriate styles for their design guidelines.

Today's residential communities need architectural variety in their streetscapes, and the implications of higher density housing and closely spaced homes may require design guidelines as well. As a method of practice, most homebuilders starting a new community will select a "product line" of at least three to five model houses that are architecturally compatible with one another.

Some of these models may offer alternate front elevations to add street view variety. The programming skill employed in this phase is critical to the design success of the new community. Ideally, the product line models in new communities will incorporate variety in mass, bulk, materials, and colors—which, when sited together on a street, will form a pleasing composition.

One of the most common problems in new single-family communities is the lack of variety in the models that are offered. Ideally, the more models the better. Some progressive builders even refuse to build the same exact model twice in the same community. But more reasonably, four to eight models with alternate façades are the norm.

Progressive builders may also integrate different housing types within the same neighborhood and street. Most new home communities are still homogeneous programs: all detached homes, all duplexes, etc. By varying building type within a neighborhood and/or street, an even greater sense of variety can be achieved that resembles communities built over time—different housing responding to different households and income levels within the same neighborhood.

The result of design controls and programming should be a vibrant community instead of a banal enclave of homogeneous architecture. Architectural controls and programming should encourage variety within common themes, suggesting a diversity of housing types and styles.

Fig. 3.18 Homebuilders starting new communities typically develop *product lines* of at least three to five models that are architecturally compatible. Chesapeake Homes' *New Port at Victory* in Portsmouth, VA had five models, each with alternative façades of varied styles found in the region

Utilizing Edges and Contrast in Community Planning

People like to live on the edge. That doesn't necessarily mean they like high speeds and danger, but people like the excitement of contrast, of differences they find when traveling to a foreign country or seeing another culture.

In residential planning and design, the concept of utilizing edges can be seen in a house next to a farm, or the desert, or a canyon. Everyone wants to buy the house that backs up to the privacy of a forest or lake—something other than just another in a row of houses. In fact, this is the source of most community-based opposition to new development—the idea that nobody wants to lose the open space that is next to them. Never mind that it belongs to someone else, private property rights, etc.— they just want to see the land remain undeveloped.

In order to capitalize on this emotion, community developers are building *open space subdivisions*, where homes are clustered on smaller lots and the perimeter land is left in a natural state, allowing the adjacent homes to back up to open space. This is similar to our earlier discussion of common vs. private open space—instead of larger individual lots that back up to one another, you have smaller lots backing up to common lands.

We designed houses for an open space community in Lancaster, Pennsylvania, called *Wheatstone*, where the original land was a farm—a controversial scenario in many parts of the country because the community loses scenic agricultural land to accommodate what is viewed by many to be tiny ticky-tacky small homes. To mitigate those damages at Wheatstone, homes were clustered on lots that were two-thirds the size of that which is normally required, with the remaining lands left

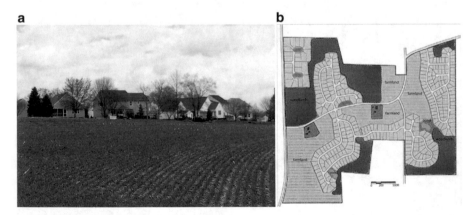

Fig. 3.19 (**a**) At Charter Homes' *Wheatstone* in Lancaster, PA, houses were clustered on smaller lots, which allows most homes to border farmland. (**b**) At *Farmview* in pastoral Bucks County, PA, Realen Homes clustered 334 homes on roughly half of a 500-acre farm tract, leaving over 180 acres under cultivation and 70 more acres as environmentally sensitive open space. (Figure (**b**) courtesy National Lands Trust)

under cultivation. Therefore, most homes would be able to back up to an "edge" of farmland offering contrast to the occupants on a permanent basis.

Realen Homes, a major builder in southeastern Pennsylvania, also built an open space subdivision of 334 home sites called *Farmview* in pastoral Bucks County, Pennsylvania. To overcome objections to the negative impact of residential development on the agricultural character of Bucks County, Realen worked with the local government to cluster the home sites on roughly half of the 500-acre tract, leaving over 180 acres of farmland and 70 acres of environmentally sensitive open space. The farmland was then deeded in perpetuity to a conservation trust.

Open space subdivisions can take many themes—wooded lands, waterfront sites, and perhaps most popular—golf course fairways. Golf course communities, in response to the market-driven interest in golf play, have become the most popular (and expensive) form of open space community for upscale homes throughout the country. Natural or man-made, expensive or free, land plans that provide home sites with edges are a beneficial and workable idea.

Building a Sense of Place

To summarize the planning and design objectives in developing new communities, one could say that the real goal is to create a *sense of place*, a feeling that people can connect to and belong to the physical environment that forms the community. The

Fig. 3.20 *Downing Woods* in Chapel Hill, NC includes houses oriented toward a permanent forest preserve

Fig. 3.21 Community open spaces with adjacent businesses that encourage gathering and socializing are important components of effective neighborhood design. The two-acre Pinehills Village Green hosts events from Art Shows to Concerts as the heart of the community and center of daily life for adjacent neighborhoods as well as residents of The Pinehills in Plymouth, MA. (Photo courtesy Pinehills, LLC)

famous comment of Gertrude Stein about Oakland, California, "there is no there there," could apply to most of our new communities as well. As we noted, most are built on stock planning and design recipes of homogeneous architecture and excessive engineering standards. By contrast, plans that introduce diversity into a new community with different housing types, architectural styles, and open space help to develop a distinctive, individual character.

Two ingredients that make a place special are (1) the physical environment, and (2) the people who live there. Certain community design goals, such as encouraging diverse housing types, can help make any neighborhood and community more interesting. In the same vein, it's tough for people within a community to interact unless there are physical places for them to meet, i.e., open space, parks, and recreation areas.

Good community design addresses these issues and tries to strike a balance between the needs for public and private space, diversity within a coherent setting, and human scale in a machine age. The goals of new community planning and design should maximize the potential for residents to develop a sense of belonging and connection to their home, neighborhood, and community.

Notes

1. "Village of Euclid v. Ambler Realty Co.", 272 U.S. 365 (1926).
2. "America in 2015: A ULI Survey of Views on Housing, Transportation, and Community". Urban Land Institute: Washington, D.C., 2015.
3. Appleyard, Donald. 1981. Livable Streets. Berkeley and Los Angeles, CA: University of California Press.
4. Ibid.
5. Randall Arendt, *Crossroads*. "Hamlet Village Town". APA Press: 1996.

Chapter 4
Siting and Lot Patterns

In this chapter, we will progress from our discussion of neighborhood and community design to a smaller scale of issues—designing housing prototypes that will fit on individual lots—yet still relate to the overall street and neighborhood.

The initial challenge of prototype housing design—selecting or developing floor plans that fit appropriately on lots—seems quite simple, but in fact it is the root cause of many problems in establishing a well-designed community. Single-family lots can be developed in a wide variety of shapes: squares, rectangles, parallelograms, triangles, etc., and lot sizes can vary from as little as 3000 sq. ft to well over 2 acres (87,120 sq. ft!). Matching house plans with the size and shape of typical lots is critical.

Complicating these programming efforts are new higher density, small-lot/house designs, mentioned briefly in Chap. 3. These "odd lots" will be discussed in detail later. Let's start first with an analysis of house designs suited for conventional lots in square or rectangular configurations.

Conventional Lots

Residential lots are typically defined by two dimensions: *frontage*, or width along the street, and *depth*, or distance from the front to the rear property line. The other key definition is *setbacks*, which determine minimum distances that structures must be set back from the property lines, as outlined in zoning regulations. Minimum yard areas, as defined by setbacks, are subtracted and the remaining portion of the lot is the buildable area, or *maximum building envelope*.

Fig. 4.1 One goal of community design is to develop housing prototypes that fit on individual lots, yet which still relate to the overall street and neighborhood

© Springer International Publishing AG 2017
J. Wentling, *Designing a Place Called Home*,
DOI 10.1007/978-3-319-47917-0_4

A generous suburban lot in the East, South or Central States, for example, might have 80 ft of frontage and 120 ft of depth. Setbacks could typically be 10 ft on either side and 30 ft in the front and rear. That leaves a buildable area of 60 ft wide by 60 ft deep, and the ground-level plan, or *footprint*, of the prototype houses must fit within this space.

In most cases, a prime consideration in selecting or designing floor plans to suit a particular lot is to take advantage of both the maximum frontage and the rear orientation. Typically, the *width* of a house (dimension parallel to the street) connotes *value*. Similarly, the privacy of a rear yard has intrinsic value that needs to be recognized in the overall lot/house design. One of the more common mistakes in the selection of plans or design process is the frequent use of plans that are too narrow for the lot. A 30-foot wide house on an 80 ft wide lot can look ridiculous, but is commonly built because someone likes a particular floor plan regardless of the lot configuration. This "small house on a big lot" situation, as seen in Fig. 2.1, also happens frequently because of large-lot zoning practices, where homebuilders have no choice but to construct smaller houses to price their product affordably. The following are some suggested rules for determining appropriate setbacks on conventional lots:

1. *Combined sideyard setbacks should generally total between 25 % and 35 % of the total lot frontage dimension at the face of the house.* Therefore, on typical lots, the distance between houses will be about 30–50 % of the typical house width. This allows houses some buffer space, yet keeps them close enough to form a coherent streetscape. Subdivisions with excessive sideyards that create distances between houses that are more than the width of the average house tend to make the houses look isolated and abandoned in relation to one another.

2. *Front setbacks should be less than rear setbacks.* Front setbacks are generally determined by alas, the distance needed for a driveway deep enough to accommodate off street parking. Front yard setbacks usually call for a minimum of 25 ft, although the front façade of the house other than the garage portion could, and often should, be closer to the street. So long as the garage is pulled back, the front of a house could be 10–15 ft from the front property line.

3. *Rear yards should be determined by whatever dimension seems to be a reasonable separation between homes for privacy purposes.* Rear yard setbacks should always be deeper than front yard setbacks—but curiously they often are not. The reason they should be more is that the front yard is set back from the front property line, which is adjacent to a street right-of-way that may be 35–50 ft wide. Houses may thus be 100 ft apart from front wall to front wall, but almost half that distance in the rear, where the need for privacy is actually greater. I recommend minimum rear yard setbacks of 30–40 ft.

4. *Reasonable front, side, and rear yards, if not called for in zoning standards, should be advocated by community developers.* Often local planning boards may be willing to change standards if sound arguments are presented. Excessive yard areas, particularly in the front and side of a home, can do major damage to the overall sense of scale within a community and should be argued against when necessary.

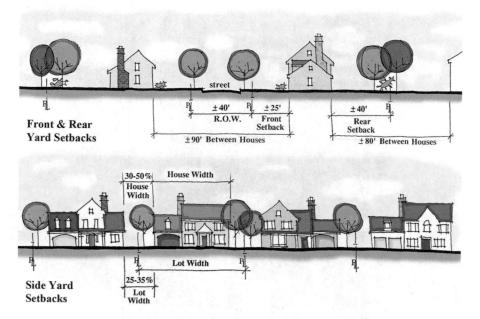

Fig. 4.2 For houses on conventional lots, the front, rear, and side yards should all be proportional to the lot and house dimensions

House/Lot Orientations

With conventional rectangular lot patterns, the two major orientations (other than for solar and environmental considerations) that a house design should relate to are, simply, the *front* and *rear* yards—which are alternatively and respectively the *public* and *private* domains of the lot and house. Sideyards function primarily as buffers between homes and are of limited use. The front and rear yards, however, are important to address in programming the floor plan, as well as outdoor spaces that will connect to and enhance the house.

Most new home designs generally recognize the value of private rear yards. Here, we typically see the key outdoor structures and amenities—decks, patios, and screened porches in the Northeast and Midwest, water features, and lanais in the South and West. It is also common to find small gardens where fruits and vegetables are grown. The rear yard is also the primary outdoor location on the lot where people engage in conversation, entertaining, or simply relaxing.

Historically, most outdoor amenities were located on the front or sideyards of the house, as recognized in designs from as recent as 25 years ago. In those prototypes, the key outdoor space was a front porch for smaller homes and a side porch for medium to larger homes. The rear of the lot was more of a utilitarian space for garages, storage sheds, hanging out the wash, etc.

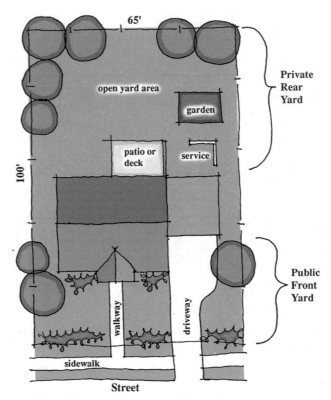

Typical Front & Rear Yard Orientation

Fig. 4.3 On conventional lots, the front and rear yards serve as the public and private domains, respectively, of the lot and ultimately the house

In effect, we have seen most of the design emphasis turn away from the front and side of houses. With the exception of the entry door, all key orientations and amenities are now focused on the rear of the lot for maximum privacy. Conversely, some functions that were in the rear were moved forward, most notably the garage. There certainly could be some social commentary here, in that housing designs seem to be acknowledging that people are becoming more antisocial, private, and internalized. One may look to the famous quote about architecture from Sir Winston Churchill: "*we shape our dwellings, then afterward our dwellings shape us.*" Clearly, housing designs need a better balance between the front and rear orientations in order to respond to the public/private roles of the home.

In some of our designs, we are recalling the side porch for large-lot houses and front porches in small-lot prototypes. In other cases, the garage is being moved back into the lot and combined with other functions to form a "service zone" shielded from the public side of the lot. The market interest in privacy must be recognized in current designs, but there is still a need to improve the public image of the house.

Fig. 4.4 Some new home designs are recalling the popularity of the side porch, as seen at *Heritage Fields* in East Lyme, CT. Locating porches on the side or front of houses improves their street appearance

Accentuated Entries

If the front façade has one important purpose, it is that of showcasing the formal entrance to the house. This is not to say it is the most commonly used entrance, but it should be the most prominent. The entry area should be gracious and inviting in appearance, for in spite of its infrequent use, it symbolizes the warmth and hospitality found inside the home. Entries and porches are places where the owner has the opportunity to decorate and personalize their home.

In spite of the symbolic importance of the entry, its design is frequently compromised in new houses. How? In some cases, it is not even visible from the street, particularly in higher density prototypes. In other cases, you can see the entry, but it is not at all prominent on the façade—or it may be in a dark or confined location, behind or between other walls.

In order to address these common problems, the following are suggested techniques to accentuate the entry area of a new house:

1. *Locate the entry in a prime location on the front façade.* This is commonly the middle of the home. It may be off-center within the front elevation but still be centered below a roof gable, porch, or other building element. It may be the most forward portion of the elevation, closest to the street.
2. *Provide a generous area in front of the entry door.* This space should allow a stoop, court or porch, of dimensions to accommodate a small group of visitors. Appropriate dimensions will vary with the size of the house and climate, but even a small affordable home should have a 5 ft wide by 4 ft deep entry stoop.

Fig. 4.5 (**a**) The formal entry should be gracious and inviting in appearance, and symbolic of the warmth and hospitality inside the home. (**b**) Unfortunately, in some higher density prototypes, the entry cannot even be seen from the street

3. *Consider providing a covered entry area.* Sheltered entries are customary on houses in some regions, but they are nice to have anywhere. Front porches tend to be more historically common on affordable houses than on larger ones, due to the everyday use of the front door on smaller houses as opposed to the more ceremonial use of them in more expensive houses. A small covering or recessed entry of some sort is recommended for new houses, particularly in areas where porches were traditionally found.

4. *Include windows or glass in or around the front door.* This simple gesture acknowledges that people like to see who they will be greeting (or not greeting) at the door, while also capturing light for the interior of the house. The most common method of adding glass to an entry area is to specify doors with integral side lights on one or both sides of the door. In other cases, glass panels may be called for right in the front door design. Still another idea is to add a small window or windows on the wall near the door.

5. *Consider including a "usable" front porch around the entry.* Based on the size and geographical location of the home, a generous front porch may be a visual and social enhancement to the house. Front porches, as mentioned earlier, tended to be found more on affordable homes than more expensive ones; however, that tradition varies by region. A usable front porch may be defined as one with dimensions of 6–8 ft in depth by 10–12 ft in width. Often they run the entire width of the front façade or even wrap around the side of the house.

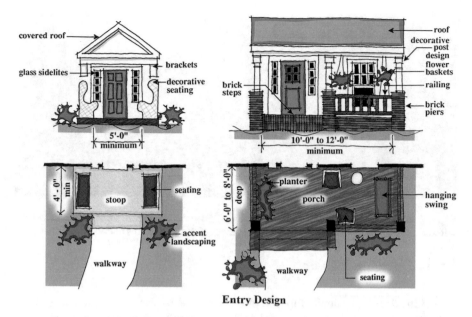

Fig. 4.6 Formal entries can be "dressed up" with amenities and human-scale details to make them inviting

6. *Add human-scale details at the entry.* Built-in seating is an example of an historical amenity seen around the front door. Planters, garden walls, columns, railings, or other features should also be considered if the budget allows. Imaginative designs for light fixtures and mailboxes on freestanding posts are often part of an inviting entry statement. Even the street numbers could be incorporated into a handsome detail that will lend scale to the entry.

7. *Use quality building materials around the entry.* Particularly with limited construction budgets, the place to loosen up the purse strings is around the entry. The stoop and steps may be brick or stone, or colored and stamped concrete at this outside location only. Porch roofs may be supported by decorative columns or wrought iron. Roofs may be an upgraded material such as copper. Railings can have decorative patterns. Walls can incorporate quality veneers around the entry areas only. The front door itself should have a pleasing design pattern.

8. *Address the need for lighting and landscaping leading to the entry.* The approach to a home must be well lit and visible for safety and street presence. Landscape material along the entry façade and walkway should frame the entry—and not hide the front door—which is all too often the case.

The formulation of a quality house design must start with a quality entry statement. A prominent and well-detailed entry design should be one of the early considerations in resolving the overall house/lot design solution.

Fig. 4.7 An older home would typically have a second entrance or back door that served as the informal, utilitarian entrance to the home, sometimes called the "friend's entrance"

Front Door/back Door

One of the features typically found in older homes of all sizes was a second entrance or back or side door. This served as the informal or utilitarian entrance to the home, with direct access to the kitchen and informal areas of the house. The back door formed a complementary role to the formal front door in most designs, with the back door used by close acquaintances—which helped coin the term of "back door guests" for the household's closest friends.

New house designs have generally eliminated the back door concept, largely due to the standard practice of attaching the garage directly to the house; the back door was transformed into an interior door between the house and garage, which is a much less gracious entry for guests. With the loss of the side door in most new home designs, the typical access to the rear yard is through a sliding door on the rear façade—which generally cannot be easily accessed from the outside.

The back door entrance is another feature that should be reincorporated into production houses where possible. Here are some suggested ways to accomplish this:

1. *Locate a "back" door on the side of the house.* The ideal place for the back door may actually not be in the back, but on the side of the home. This placement allows comfortable movement from the front or rear or the house, allowing occupants more opportunities for easy access to the informal areas of the house.
2. *Locate the back door near the garage, but not in the garage.* The back door should be near the automobile storage for easy access, and not hidden within an enclosed garage where it would have restricted access. This location will preclude the need to go through the garage to use the back door.
3. *Consider a breezeway between the garage and the house.* One potential solution to incorporate a back door is to connect the garage to the house with an open breezeway. The breezeway allows a direct covered path from the garage to the house, with access from both the front and rear yards. In some cases, the breezeway may be large enough to become an open-air screened porch or

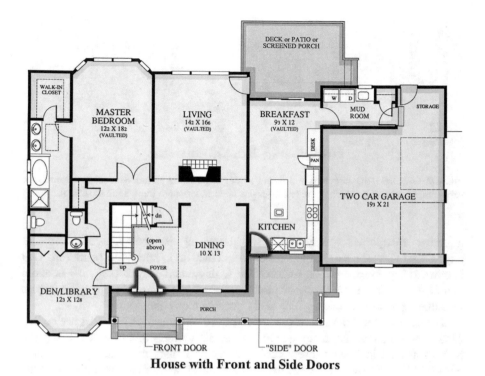

House with Front and Side Doors

Fig. 4.8 Floor plans for new houses should reincorporate a second entry location, generally near a kitchen or informal area of the house as a "friend's entry" for guests or simply as a convenience

even an enclosed room between the garage and the house—both popular home improvement projects.

4. *Use a hinged door along the rear façade as the back door.* If the house design is very small for affordability reasons, a hinged swing door of some sort that can be opened from the outside (instead of a sliding door which cannot) can be located in one of the rear-oriented rooms, such as the kitchen or living/family room.

The inclusion of a usable and accessible side or back door, coordinated with the more formal front entry door, will help make new house designs more functional as well as livable. Access to the outdoor environment should be made easy and convenient for occupants.

Accommodating Autos

One of the most important issues to resolve in developing a positive house/lot relationship in new houses is usually designating an appropriate location for the parking and storage of automobiles. In most cases, this means siting a driveway that leads to

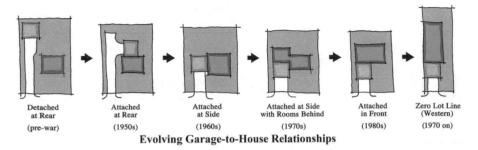

| Detached at Rear | Attached at Rear | Attached at Side | Attached at Side with Rooms Behind | Attached in Front | Zero Lot Line (Western) |
| (pre-war) | (1950s) | (1960s) | (1970s) | (1980s) | (1970 on) |

Evolving Garage-to-House Relationships

Fig. 4.9 Over time, the garage moved from the rear of the lot as an independent structure to the very front of the house, which is the prime reason most streetscapes today look so dismal

a garage or carport. You may have heard the saying that the average homeowner pays more to house his or her automobile than they do to house their children. That is correct. The average space required for a single car—say 240 sq. ft—is about double the size of an average secondary bedroom. What's more, the garage occupies valuable ground floor space, which is at a premium on small lots.

The saga of deterioration of garage-to-house relationships is worth reviewing. Most prewar homes had the garage located as a separate structure in the rear yard. Market interest in having direct access from the house to the garage prompted homebuilders to move the garage forward and attach it—initially to the back of the house. In an effort to reduce the length of driveways and to open up rear yards, the garage was then moved to the side of the house, with the garage doors facing the side of the lot on larger lots or facing the street on smaller lots.

With the garage typically being found on the side of the home, many plans began to introduce rooms behind the garage—which required that the garage be moved in front of the entry façade of the house. In many current designs, the garage is tacked on to the front of the house to allow for maximum flexibility in the interior floor plan.

On many small or narrow lots, the garage location issue boils down to one question: which do you want to compromise, the interior plan or the exterior appearance? With the garage pulled way out in front of the house, and the entrance pulled way back, the floor plan can minimize circulation and construction costs. And as discussed earlier, houses are merchandized to sell from the inside, so most often the exterior appearance of the house loses out to interior efficiency. This is a prime reason for why our streetscapes look so bad!

Let's analyze the garage placement issue by first addressing the most common garage location: on the side of the house in a *front-loaded* position. In that case, the following rules can mitigate the visual impact of the garage:

1. *The front wall of the garage should not be in front of the entry façade of the house.* To follow this rule, you need a footprint width of at least 40 ft for a one-car garage. A two-car garage would require more like 50 ft, and a three-car garage 60 ft. Footprint widths below those minimums will probably require a frontal garage arrangement that results in an "odd lot" house/lot orientation.

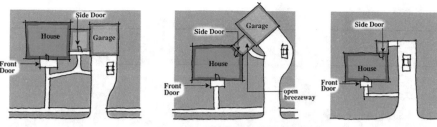

Front and Side Door Issues

Fig. 4.10 The second entry or "side door" feature should be located for gracious access into the home—which means *not* going through the garage. A breezeway or porch may provide a better location for this door

2. *Minimize the width of the garage.* Beyond functional needs, garages should have the minimum dimension parallel to the street, to allow the front elevation to be dominated by the habitable portion of the house. This role means that places for storage are better located in the rear of a front-loaded garage, such that the garage width is only determined by vehicle dimensions. I generally recommend that the minimum width be 12 ft for one car, 20 for 2, and 30 for 3. In affordable houses, these dimensions can be reduced to 10 ft for one car and 18 ft for two cars.

3. *Only provide garage spaces for the number of cars needed per market standards.* If a market survey indicates that a household typically has one car, there's no need for a two-car garage, particularly when lots are small and two-car garages tend to damage the street view of the house. *For affordable programs, consider offering a one-and-a-half car garage,* which will allow a single car and some extra storage space. These can be about 14 ft wide. If houses are presold (sold before construction), designs can be offered with a "one-car garage standard, two-car garage optional" as the buyer's choice. The same concept works for two-car garages standard, three-car garages optional.

4. *If you can achieve a side garage placement, you can then "side load" the garage.* For homes built on corner or oversize lots, side-loaded garages are standard practice. For more narrow lots, some homebuilders are reconsidering ideas such as the shared driveway, which can compensate for the fact that additional paving is required for side or rear loading. However, designs must consider and eliminate the potential for blockage of the driveway by neighbors and shared maintenance—factors which are major market objections to the shared driveway.

5. *Where front-loaded garages are necessary, pull the garage back from the front façade as much as possible.* When garages are along the side of a house, recess the garage into the lot as much as is reasonable (4–12 ft) considering the interior plan of the house and exterior lot conditions. This may mean altering the interior layout such that there is more circulation or less desirable access from the garage, which is part of the process of balancing interior and exterior design concerns.

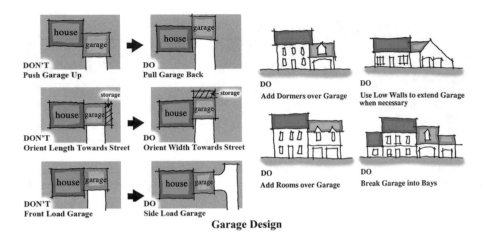

Garage Design

Fig. 4.11 The side-located, attached garage is the most common orientation found on conventional lots. There are many design techniques that can be used to mitigate an overbearing presence in this position

6. *Pull garage walls low to the ground.* Garage walls often need not be the same height as walls around interior space. In portions of the garage used primarily for storage, consider reducing the height of the walls, so that they are closer to the ground and less prominent, such as between 4 and 6 ft.

7. *For compact two-story plans, consider rooms over the garage.* This integrates the garage into the house design and pulls the eye away from the ground-level garage when viewing the house from the street. For small houses, it may also be a more cost effective way to deliver square footage.

8. *Use quality materials in and around the garage doors.* Pronounced or arched trim molding around garage doors and specifying doors with integral panels and windows is also recommended for front-loaded garages. When garage doors are in full view from the front elevation, a well-detailed appearance is critical.

9. *Consider the impact of color selection of the garage door.* Lighter colors that blend into the rest of the house façade or make the garage door recede are suggested. In general, use the same colors for the house on garage walls and doors. Using vibrant or strong colors at the garage is a common, yet major, mistake.

10. *Three-car garages should have breaks in the wall and rooflines.* Three-car garages should be broken into at least two masses by pulling the wall of one bay back into the building envelope. In some designs, the garage spaces are pulled completely apart and located at either side of the lot. These and other special plan arrangements should be considered for mitigating the scale of three-car garages at the street.

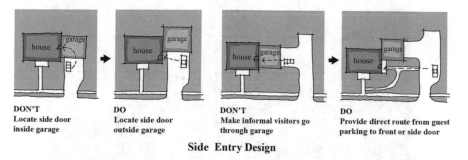

DON'T
Locate side door
inside garage

DO
Locate side door
outside garage

DON'T
Make informal visitors go
through garage

DO
Provide direct route from guest
parking to front or side door

Side Entry Design

Fig. 4.12 People will naturally use the most direct route from their car into the house. Don't make visitors go through a junky garage to get inside!

Later, I will discuss other alternatives to the side-located, front-loaded garage, including rear-located garages, alleys, and other ideas for special narrow-lot conditions. The most common placement of the garage, however, is at the side of the house—a practice that often could be better resolved by considering the above suggestions.

Driveways, Parking, and Doors

Once the homeowner has parked his or her car in the garage, they typically gain direct access to the informal area of the home through what used to be the back door, which is now generally the "garage door." Now the question is—how do other people, i.e., guests and visitors, enter the house—is there a good formal entrance that coordinates with the street, sidewalk, and parking locations? Is an informal entrance such as a back door possible?

As a minimum, site/lot plans should allow for at least one secondary parking place other than the garage, and this additional parking should be available on site, or on the street near the house. From the parking location, the front entry door should be visually prominent and easily accessible by walkways, and the back door (if included) secondarily visible. Visitors should have no question as to where these doors are, which is which, and how to get to them.

Unfortunately, side-loaded garages frequently present a problem for guests when parking spaces are located such that the closest access to the home is *through* the garage, which is a most ungracious entry, among other things. Designers realize that people, subject to human nature, will take the most direct route between two points, which means they will go through the garage unless an alternative is presented. A carefully located back door, when coupled with clear pathways and smart landscaping, should direct formal and informal visitors away from the garage when entering the house.

For pedestrians, walkways leading to the entrances of the house should also relate to the sidewalk (if any) and street. Ideally, visitors parking in the driveway or street and pedestrians should have a clearly recognizable route to the entry door(s).

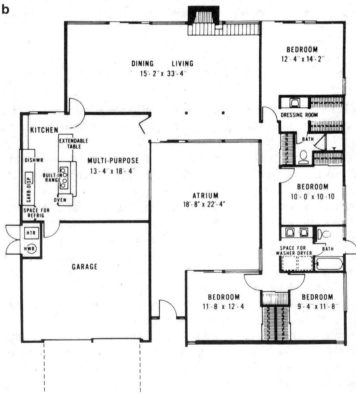

Fig. 4.13 San Francisco-based builder Joseph Eichler developed (**a**) small-lot houses with internal atriums. (**b**) Eichler's mid-1950s houses averaged 1200 sq. ft and were built on 7000-square-foot lots. Later, zero-lot-line houses would become much larger, while lot sizes themselves shrank. (Fig. (**a**) courtesy Ernest Braun Photography. Fig. (**b**) courtesy Stantec, formerly Anshen & Allen Architects)

Small Lots and "Odd Lots"

If the available footprint cannot accommodate a side-located garage, or if the lot size becomes very small to achieve higher density objectives, a small-lot house design becomes necessary. As discussed earlier, California and many Western states have a severe problem exactly opposite that of the East and Midwest—lot sizes have become excessively small. How small? As mentioned in Chap. 2, in California you can easily find a 2500 sq. ft house with a three-car garage on a 3000 sq. ft lot!

In terms of the historical background of Western small-lot houses, the early Mission courtyard plans established precedent for later patio and atrium homes oriented toward the comfortable outdoors. Some of the best examples of production *atrium houses* were those built by San Francisco builder Joseph Eichler between 1947 and 1963. In the mid-1950s, Eichler's houses averaged 1200 sq. ft, built on 7000 sq. ft lots and priced at $11,950! Eichler's atrium homes had blank walls around the perimeter with most rooms focused on a private interior patio.

Early *zero-lot-line houses*, which were loosely based on atrium/courtyard plans, were introduced in the 1960s, not necessarily to increase density, but to make better use of the surrounding yards. Aggregating two sideyards on one side of the house created three usable yards (front, rear, and side) instead of two. This newly enlarged sideyard generally abutted a blank wall on the adjacent home, which was typically located directly on or near the property line.

In the 1970s and 1980s, residential designers continued to favor the zero-lot-line approach, which became highly problematic when lot frontages dropped below 50 ft and two-car garages became commonplace. At this point, designers were forced to place the garage totally in front of the house and locate the entry at the mid-point of the interior floor plan, not visible from the street. A collection of these often resulting in a devastatingly poor street scene.

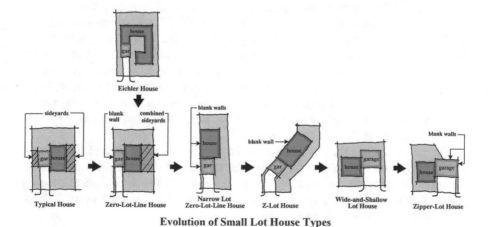

Fig. 4.14 Western small-lot houses evolved from courtyard designs to zero-lot-line and various other house/lot configurations during the 1970s and 1980s

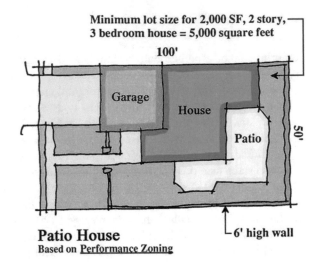

Patio House
Based on <u>Performance Zoning</u>

Fig. 4.15 In his classic 1980 book, <u>Performance Zoning</u>, planner Lane Kendig developed a formula to limit building intensity based on the carrying capacity of the land, as shown above. Most small-lot houses in Western states effectively doubled Kendig's recommended housing size standards

In the early 1980s, homebuilders and designers began to develop alternate solutions to the narrow-lot prototype. One initial concept was the *Z-lot*—and later, the *angled Z-lot*. Both schemes configured lot lines so that the home would face the street on an angle to allow more of the front entry façade to be viewed from the street. In theory, this meant that garage doors should be visually less prominent on the façade since they were not perpendicular to the street.

The Z-lot schemes were an improvement over standard narrow-lot designs, but still left a lot to be desired. When viewing a street of Z-lot houses from certain angles, the Z-lots actually made the aggregation of garage doors seem worse. As an alternative solution to this dilemma, some residential designers began to propose another idea—*wide-and-shallow lots*. This lot was considered a radical new concept by the home building industry. The idea behind wide-and-shallow lots (to simultaneously widen street frontage and reduce depth) was not new at all. They were common on many prewar subdivisions, but those designs did not include two- and three-car garages at the side of the home.

The new wide-and-shallow lots were a hit with Western buyers and the housing industry press, but there were still some design negatives that needed to be addressed. The primary drawback of these lots was the diminished rear yard, considered a prime amenity by homebuyers. Rear yards were sometimes as little as 10 ft deep, which only left about 20 ft between houses.

In response to this problem, blank walls similar to the type used in zero-lot-line homes were incorporated into wide-and-shallow houses. These blank walls were later combined with jogged rear lot lines to form *zipper lots*, named for the pattern created by the back-and-forth movement of the rear lot lines.

While each of the Western odd lot patterns has its positives and negatives, the overriding problem is basic: the lots are too small for the size of house being built on them. Western states now have millions of houses like the ones you see in Spielberg films, garage doors wide open displaying messy junk to the community, making for a dismal neighborhood presence. However, these are still being built by the thousands each year, with little practice of new alternatives.

Is there a solution to the Western small-lot problem? No. The dilemma goes straight to some of the toughest issues that developers and planners face, especially in highly desirable locales such as California. The solution lies between abandoning rational design standards for major density increases, and creating elitist communities. Unfortunately, the benefits of increased density are short-lived.

While the initial density increases made possible by smaller lots were of some help in addressing affordability, the laws of supply and demand will always continue to force prices higher. Now, very small lots, such as 3000 sq. ft—have become the norm for even upscale 3000 sq. ft houses. Although downzoning land to require larger lots is possible, it is highly unlikely given the political consequences.

With respect to the issue of lot sizes in Western markets, I like to refer to the work of planner Lane Kendig in his book, Performance Zoning. My problem with small lots does not preclude the use of 3000 sq. ft lots, but I do object when 3000 sq. ft houses are placed on them. In Performance Zoning, Kendig develops standards for medium- to high-density housing types based on impervious surface coverage, floor-area-ratios, spacing between houses, and other physical standards. Kendig recommends floor-area-ratios (the ratio of heated interior space to lot area) for 3-bedroom, high-density detached houses, for example, to be 0.33 for *Patio Homes*, and 0.48 for *Atrium Homes*, the highest density category.[1] This limits the size of houses placed on 3000 sq. ft lots to around 1500 sq. ft of interior heated space. Conversely, a 2000 sq. ft house would need at least a 5000 sq. ft lot. The existing small-lot house designs that are considered the norm in Western states effectively double Kendig's recommended housing size standards.

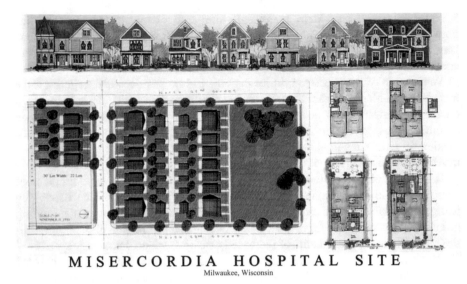

MISERCORDIA HOSPITAL SITE
Milwaukee, Wisconsin

Fig. 4.16 Locating the garage in the rear of the lot, with access from an alley or lane, is a popular concept among traditionalist planners and based on historical precedents found in prewar communities

Fig. 4.17 The *New Neighborhood in Old Davidson*, in Davidson, NC, had accessory units above garages

There have been some progressive designs in response to the small-lot issue in Western housing markets. One track I see is traditionalist, and the other is Europeanist. The traditionalists, as discussed in Chap. 2, experiment with planning and design criteria that depart from local standards and reintroduce historical concepts like alleys and rear garage placements.

Laguna West was the first of Calthorpe's "Pedestrian Pockets," or "Transit-Oriented-Developments" (TODs) designed such that residents can walk anywhere within the community in 15 min and have access via public transportation to the greater metropolitan area. Calthorpe's program for a Pedestrian Pocket would include office, retail, parks, and civic facilities, complemented by an appropriate mix of different types of housing—all linked to a larger city by mass-transit. To achieve this, Calthorpe recommends 120–150 acres of land, though smaller versions can be built with reduced acreage. Unlike *Seaside*, the Pedestrian Pocket model does not imply that traditional planning principles be used to implement the design, even though programmatically it is based on emulating a village scale.

While a community like *Seaside* provides a long-term model for community design, Europeanists address the short-term resolutions to the small-lot dilemma with the following principle—if we are going to live at higher densities, let's model our communities on the ancient cities of Europe—only somehow add cars. With the Europeanist approach, we see an extremely high degree of finish and landscaping from the street. Houses are designed to have cars put away in their garages or not be on the street at all. Driveways are minimized so that cars can't be left outside and other parking spaces are located in guest parking areas away from the front of the house.

Fig. 4.18 A more European design approach is seen at *Bellagio* in Calabasas, CA, where we see a high degree of finish from the street. Houses are designed to have cars put away in garages or not be on the street at all. Here, a second-level loggia gives life to the public realm. (Courtesy Berkus Group Architects)

Some Europeanist models elevate living spaces above the street, architecturally expressed in the form of articulated loggias and windows that give life to the public realm. Again, very high-quality finish materials and carefully orchestrated design (at a relatively high cost) is a hallmark of this approach, thus it is difficult to achieve with affordable housing.

Narrow-Lot Design: Entry Area

Although the traditionalists and the Europeanists are inspirational, most builders must still contend with the day-to-day limitations dictated by their competition—and still need to build on "plain vanilla" narrow lots within tight construction budgets. Narrow lots may typically be between 35 ft and 50 ft wide, and from 85 to 110 ft deep. Due to the shallow width of the lot, houses typically have one or more blank walls and extensive privacy fencing. Unlike the front/back orientations of "conventional" lots, narrow-lot houses have to consider organizing the plan to front, side, and rear yard areas.

As with conventional lot houses, narrow-lot designs have important rooms that are rear oriented although sometimes the side or courtyard orientation is more favorable. Many narrow-lot homes are single story, are in mild climates, and have blank walls limiting window locations, so entrances to narrow-lot houses are pulled back to the interior of the plan. This is an important distinction between narrow and conventional lot houses and can present major problems from a community design point of view: on many narrow-lot houses, the entry door is not visible from the street, confusing visitors as to how to approach the house. If the front door is not visible from the street, then design techniques must be used to help identify the front door location.

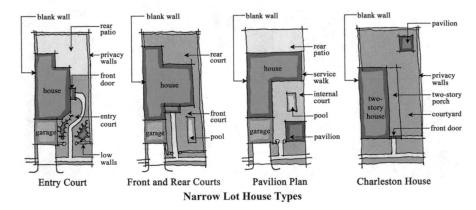

Narrow Lot House Types

Fig. 4.19 Narrow-lot house prototypes should treat their entry courtyards in a fashion similar to the public/private relationships of front and year yards on conventional lots

With a rational design approach, front and sideyard areas can be developed as courtyards in narrow-lot housing—maximizing access between the inside and outside of the house, much in the way the early Missions were organized. Many lots do not allow for this because setbacks and open space on the lot is so minimal that there simply isn't any room for courtyards.

The following are some suggestions to design narrow-lot houses to enhance street presence and courtyard living:

1. *Use an entry court to develop an entry sequence from the street.* Given that most narrow-lot houses need to locate the entry door central to the floor plan, it follows then that designs should incorporate architectural and landscape treatments to direct visitors along a path to the entry point. This is frequently accomplished with using paving materials, garden walls, gates, landscaping, and lighting elements that clearly lead the visitor to the front door. An increasingly popular way for builders to visually identify the entry point of narrow-lot houses from the street is using a raised pergola along the front wall of the entry court that then leads to a courtyard pathway connecting to the front door.

2. *Maximize the appeal of entry courts.* Narrow-lot designs should capitalize on their de facto entry courts by developing semiprivate outdoor areas that function in the manner of historic front porches. Garden walls should be low enough to see over but high enough to clearly define the boundaries of the courtyard. Some narrow-lot designs address the entry court by making it the primary yard area—even locating a water feature, garden, or other amenity there.

3. *Consider the use of a "pavilion" floor plan.* Some progressive designs are totally detaching some rooms from the rest of the house—and turning them into separate pavilions at the front of the lot to enhance privacy and access to the outdoors. This idea of breaking the home into parts works well with the trend toward home offices, private guest suites, and a remote children's zone.

Fig. 4.20 This inviting entrance to a narrow-lot house at *Leisure Village* in Oceanside, CA uses garden walls and pergolas to lead visitors through an entry sequence. (Courtesy Berkus Group Architects)

4. *Consider the benefit of a two-story plan to enhance a ground-level courtyard.* Small-lot ranch plans consume precious ground area that could be devoted to larger, more livable courtyards. Buyers who prefer single-level living may accept having lesser-used rooms in a second level—up or down—from the main floor.

The historical sideyard houses of Charleston, South Carolina offer some ideas for narrow-lot house design. Within their narrow façades, Charleston houses still had a front door facing the street—actually a cross between a door and a gate, however, since it opened to a private, outdoor court—from which the interior front door was then reached. Charleston houses typically had full-length porches on both of their two floors, which maximized the orientation of interior rooms to the courtyard.

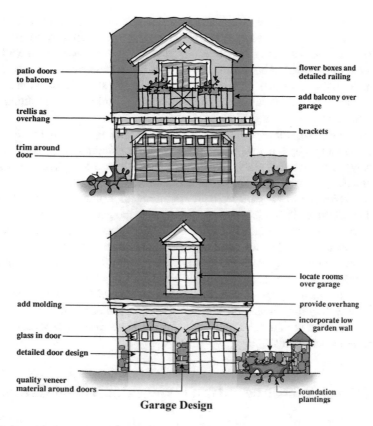

Fig. 4.21 Designing a narrow-lot house with a frontal garage presents challenges for even the most seasoned designer. Quality materials and details surrounding the door and habitable spaces above can help

Narrow-Lot Houses with Frontal Garages

Narrow-lot housing designs must respond to the fact that a major portion of the front elevation, sometimes as much as 50 % or more, will be garage doors. It's given the garage is the most forward and prominent element of the house. This situation presents challenges for even the most seasoned residential designer. The following are some ideas that can be considered to mitigate this harsh reality:

1. *The garage door design and materials should be of high quality.* This means specifying a pattern that will break up the mass of what can be a 16 ft by 8 ft element. Windows integrated into door panels always help add scale and a sense of habitation to an otherwise dead space. Raised panels or trim also add a finer level of detail to a large mass.
2. *Use two single-car garage doors instead of one double door.* When possible, use two eight-foot or nine-foot wide garage doors instead of one 16 ft

"double-car" door. In three-car garages, try to use at least two. And maybe three. Single doors, and jog the wall of the garage between doors.

3. *Use high-quality veneer materials around the garage door.* If the wall around the garage is the most prominent part of the facade, why not cover it with high-quality finishes? This is a good place to add some brick, stone, or other quality veneer material. Also, consider adding a unique trim molding such as a curved header or angled corners around the door.

4. *Pull the garage doors back from the front wall of the garage.* Recess the garage doors by one to two feet. This creates a "shadow line" that will downplay the presence of the garage door on the façade. An applied overhang such as a trellis or pent roof above the door can also be effective.

5. *Build rooms over the garage.* In programming floor plans for narrow-lot houses, it is helpful to locate second floor rooms over the garage. This will provide the entry façade with windows facing the street; they can in turn be dressed up with planter boxes, balconies, loggias, etc. Rooms over the garage will also help move the eye away from the garage doors below. The second floor may even cantilever over the garage to provide a shadow line over the doors.

6. *Incorporate low garden walls into the garage façade.* Consider extending the front wall of the garage into a low garden wall. This garden wall can be highly articulated and add scale to the street with flower boxes, decorative material patterns, and landscaping.

Other Small-Lot Garage Ideas: Garage Courts and Carports

One popular concept for accommodating the automobile on small lots is the *garage court*. With this idea, the garage is pulled back toward the rear of the lot so cars will not block the front elevation of the house. In order to compensate for the land area that garage courts require, higher quality paving materials can be used—and the driveway can double as a patio or court area. Many outdoor activities naturally take place on paved surfaces such as driveways anyway, so why cover the lot with more paving than you need to?

With the garage court concept, the front façade of the home can be pulled closer to the street and possibly include a front porch or other details to enliven the streetscape. The driveway area can be further defined with garden walls and may even include a gated entrance.

The garage court has some characteristics of the historical open-air *carport*, seen off to the side of the house in older designs. The carport was used in lieu of a garage in more affordable designs, since they were less expensive to build. Essentially, they are just open garages, providing a covered roof over the automobile, but no sidewalls. Carports have been used in some recent progressive designs, though I feel they still present problems in that they often lack the auxiliary storage space an enclosed garage generally provides.

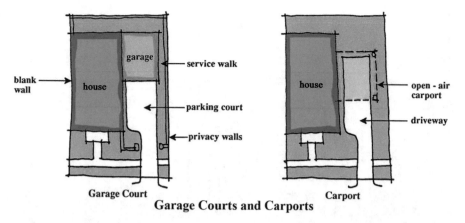

Garage Courts and Carports

Fig. 4.22 Garage courts and carports are examples of other potential small-lot garage resolutions. In both solutions, the car is pulled away from the front of the lot

Tandem Garages

Another solution being explored in the markets demanding three-car garages are in demand is the *tandem garage*, where one car is parked in front of the other. The obvious drawback of this idea is inconvenience since one of the cars is "parked-in" by another. But this negative must be considered as a trade-off, where the benefit is reducing the presence of garage doors on the façade.

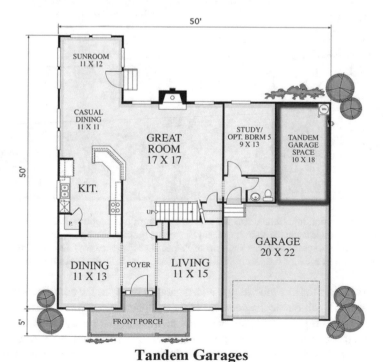

Tandem Garages

Fig. 4.23 This plan for a Virginia homebuilder provided a third parking space in the garage although it would be blocked in. In spite of the inconvenience, the improved streetscape is a valid trade-off

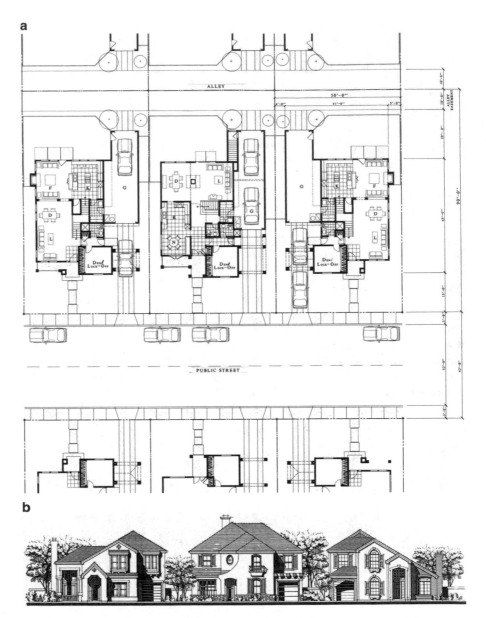

Fig. 4.24 Tandem parking—where cars are parked in front of one another—is an idea whose time has come. (**a**) In this imaginative concept, Aram Bassenian proposes that garages be served by both front and rear driveways to avoid the problem of "blocking in" parked cars. (**b**) The street view of the houses is thereby relieved from additional garage doors. (Courtesy Bassenian/Lagoni Architects)

One novel idea for implementing tandem garages includes the combination of front and rear access to eliminate the negative of being "parked-in." This scheme combines a rear alley with a typical front yard driveway to allow both a front and rear door to the garage. In other cases, the blocked space would still appeal to smaller households that would have some—though limited—need for a third space. Even for buyers with only two cars, additional space in the garage is almost always desirable for storage. Tandem garages offer a reasonable solution to reducing the impact of garage doors on the street view of the house, particularly in higher density communities.

Turned Garages

Another design concept for accommodating garages on narrow lots as well as not-so-narrow lots is the idea of the *turned garage*. Here, the garage is pulled all the way to the front of the house and entered from the side, so the driveway curves 90° as it enters the lot. On wide lot houses, this idea has been popular with upscale buyers because it locates the driveway in front of the entrance so that a gracious entry sequence is possible. (The turned garage is immensely popular in Beverly Hills, California and South Florida.) On narrow lots, however, the turned garage idea presents some challenges.

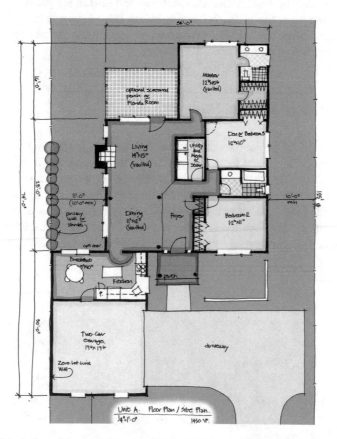

Fig. 4.25 This design with a turned garage orients the driveway and garage doors away from the street. This concept is popular for narrow lots, but parked cars can potentially block the view of the entry area

Turned garages are theoretically possible on lots as narrow as 50 ft, but those dimensions make for a tight fit. Assuming the garage is 20 ft wide, this leaves only 30 ft for backing up and turning, along with a sideyard buffer, if any. Consequently, turned garages generally need lot sizes over 50 ft, and at that point, it may be better to just move the garage back into the lot.

The other problem with the turned garage concept on narrow lots is that the secondary parking space in front of the garage ends up blocking the view of the house—and may cramp the entry area as well. Care must be taken to allow 10 ft or more from the driveway to the front door of the house, which will further remove the entry from the street view.

A turned garage, combined with some type of angled lot scheme, actually works best to solve many of these negatives. Once lots and homes are put on an angle less than perpendicular to the street, the turning radius issue becomes less of a problem. And if a shared driveway between two turned garage homes can be implemented, it can reduce the risk of closely spaced driveways.

Alleys and Rear-Loaded Garages

Yet another solution to the garage location on a narrow lot (or any size lot) is the idea of entering the garage from the rear of the lot via an alley or secondary private street. This idea was quite popular earlier in the century and remains very much in use in some parts of the southwest, notably Texas. Although the rear alley is arguably the best solution to the garage placement issue, it has been resisted by many community developers due to the increased costs of additional paving and maintenance. Some planning officials even worry about increased crime from alleys.

Fig. 4.26 (**a**) In the traditionally planned community of Celebration, FL, most garages are removed from the front and relocated in rear alleyways making for a more pleasing street scene (**b**). (Celebration, FL)

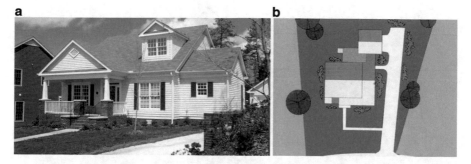

Fig. 4.27 (**a**) At the *McConnell* community, new homes have detached rear garages allowing generous porches to dominate the front façades. (**b**) Builders are reconsidering the historical location of the garage: in the back of the lot and configured to be either attached or detached

Other problems with alleys: (1) they may add more impervious area to the site, (2) they consume land such that densities are no more than conventional housing prototypes, and (3) they can be a nuisance area in terms of traffic blockage, trash pick-up, etc. Probably, the biggest reason for their lack of recent popularity, however, is the fact that the prime rear yard location might be viewed as compromised by the intrusion of garages and semipublic roads.

Nevertheless, alleyways are regaining popularity. Many traditionalist planners are reincorporating alleys into their new communities. From a community design point of view, alleys make a lot of sense—they get the cars completely away from the front of the home! Other service functions, such as garbage pick-up, can, and should, be located in the back of the house away from the street, allowing front yards to be totally free of driveways.

Rear-Located, Attached, or Detached Garages

Both community design concerns and affordability issues are motivating homebuilders to reconsider the historical location of the garage in the back of the lot in either an attached or detached configuration. This solution is very practical in that it creates a "service" zone for cars, workshops, trash storage, etc. that is out of immediate view from the street.

The *detached rear garage* can raise some marketing concerns, some of which were mentioned earlier—namely that most people prefer to have direct access from their garage into their house. These concerns would lead to the conclusion that in higher pricing ranges a garage location that allows a connection between garage and house would be more preferable. A recently established method to accomplish this connection locates the garage in back of the house on an angle (such as 45°) from the driveway.

Another solution is to site the garage in the rear of the lot independent of the structure of the house, but connect the garage to the home with a breezeway or covered walkway. In either instance, care must be taken not to compromise the value of the rear lot, which is still highly regarded as the prime outdoor amenity area. With

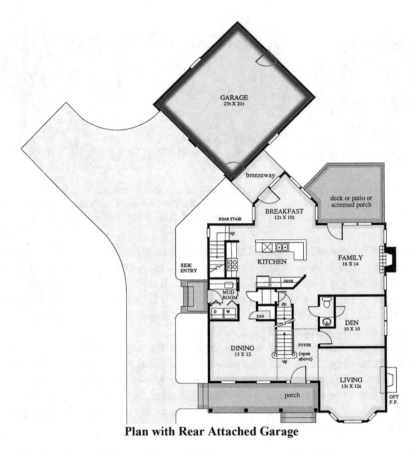

Plan with Rear Attached Garage

Fig. 4.28 Some house designs at *Summerfield* have garages angled at 45° to a rear corner of the house to pull the garage back from the street while still keeping it attached to the house

careful site planning, a "public" rear yard area can be screened from the private "service" zone for work and storage.

In more affordable houses, the idea of a detached rear garage is also being reconsidered for several reasons. First, it can easily be built at a later date without impacting the initial design or cost of the house. Secondly, on narrow lots, the location of a two-car garage in the rear requires less lot width—and the front façade will look better without a garage.

Lower Level Garages

On sites that have a grade differential between the street and the rear of the lot, one could consider the *lower level garage* location. With this scenario, the driveway descends a full level to be flush with the basement, under the main living floor of the house. The lower level garage was historically popular when land was less

Fig. 4.29 A lower level garage, seen in this design for Chesapeake Homes in Raleigh, NC, offers another concept for removing the garage from the front façade. Lower level garages can also reduce construction costs by integrating the garage's structure into the house

expensive because it makes the house more efficient structurally. A lower garage typically needs a larger lot than a frontal garage; therefore, it is infrequently used in new small-lot designs, in spite of the structural savings.

For larger and more upscale houses, the lower garage is still a viable idea to enhance front elevations. And with the increased popularity of three-car garages, the lower level garage warrants additional consideration as an opportunity to rid the front elevation of garage doors altogether, and possibly save on structural costs in the process. A more compact footprint and integrated structure result in having a garage below, and construction costs lower.

A marketing consideration with lower level garages is the relationship of the garage to the kitchen, specifically the capacity to minimize the task of climbing stairs with groceries. This can be mitigated with an easily accessible, comfortable stair that runs directly up to the kitchen and functional areas of the main floor. Often this stair will need to be a "second" or even "third" stair, such that the main stair location is not compromised and the access path from the garage follows the most direct route.

Community and Privacy in Small Lots

All types of lots and all types of houses should address the balance between community and privacy. Earlier, I discussed the rather simple front/back relationships of conventional lots. Higher density housing on small lots faces the same issues and needs to develop alternative outdoor orientation. Ideally, each lot/house pattern should provide for connecting to the larger community in one area as well as retreating from it in another.

In the next chapter, I will begin to discuss the logistics of interior room arrangements, which also concerns linkages between indoor/outdoor spaces. Certain interior rooms may favor a more public location, while others should be privately oriented. But before moving inside the house, I would like to first discuss exterior areas around the house and how to resolve designs for these "outdoor rooms" for both community and privacy issues.

Porches, Patios, and Decks

As previously bemoaned, the key outdoor areas on historical designs—generally covered porches on the front or side of homes—are now in the back of homes as rear decks or patios. Typical rear outdoor spaces range from tiny concrete patio slabs to elaborate wooden decks and water features with lanais (verandas), while the fronts of new houses now have only small decorative porches or modest stoops.

Rear decks have gained immense popularity in postwar housing. They've replaced what used to be the outdoor covered porch simply because in the postwar era, decks have been cheaper to build than covered porches. In some markets, builders even omit decks from their finished products and let the homeowner build it themselves as a home improvement project to lower the base price of the house and allow individual customization of the deck.

Decks can very specifically define outdoor spaces: they can be left open to the sky, covered with a roof, or enclosed with insect screens or even glass panels to create a seasonal sunroom space. Since decks and porches are outside, however, designs are generally not given much attention and often fail to address the basic

Fig. 4.30 The historic outdoor porch has been repurposed into the oft-unsightly rear deck

considerations given to interior rooms—such as furniture placement, circulation, and clearances. Yet, as outdoor rooms, decks clearly have the potential to be the most delightful place "in" the house.

By far the most common problem with decks or porches in production homes is that they are simply too small to be usable and are not private enough to be enjoyed (often they are so high above grade people feel like they are on a stage). This problem is more common in the small porches included for effect only in front of the houses, rather than in the back; however, both locations can end up ridiculously small and useless. Rear patios and decks should be at least 10 ft deep by 12 ft wide, while front porches should be at least 6 ft wide and preferably 8 or 10 ft wide, if intended to be primary usable outdoor areas.

Deck/porch design offers a big opportunity to make production housing more marketable and livable. Historical examples of porch amenities include built-in seating, hanging swings, flower boxes, planters, and detailed railings—which can give today's builders quite a few ideas for options that can be offered prior to sale.

1. *Design decks for maximum air circulation.* Porches or decks located such that they are open on three sides are optimal for ventilation and sun access. Therefore, design porches as projections from exterior walls if possible. Patios or decks with two or more sides connected to walls of the house lose their open-air quality.

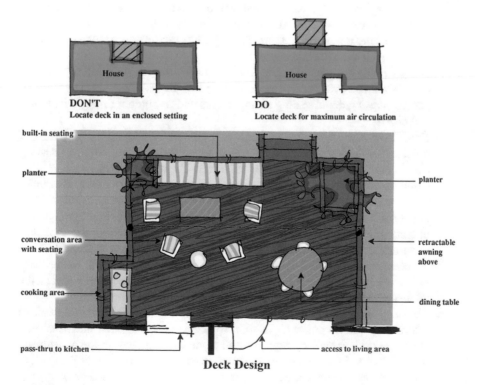

Deck Design

Fig. 4.31 Decks, patios, and porches should be designed with the same care taken as with an indoor room: they must be large enough to accommodate furniture, circulation, dining, and other social functions

Fig. 4.32 A deck or patio should be designed to be ready for an optional upgrade to a screened porch—another amenity popular with buyers, particularly in warmer climates

2. *Locate access to decks from frequently used interior rooms.* Deck locations should relate as much as possible to the more heavily used rooms of the house—typically the kitchen, breakfast, and living spaces. Decks and patios can also serve as an intermediate space between the house interior and the balance of the lot, so a direct circulation path from the house to the yard needs to be considered in deck designs.

3. *Design decks to accommodate outdoor furniture.* As with interior rooms, decks should comfortably provide seating for groups of people. They should permit conversational seating arrangements and/or outdoor dining with tables and chairs.

4. *Consider options for built-in furniture.* A linear bench along the perimeter of a deck can be designed so that it also functions as a rail. Built-in seating should anticipate and complement moveable chairs and tables.

5. *Provide places for outdoor cooking on decks.* Patios or decks should also anticipate the need for outdoor cooking in a comfortable and gracious location. This, of course, means the barbecue grill needs a decent spot. Older homes frequently upgraded the outdoor cooking amenity to a masonry fireplace, which makes sense and is now seen in brick outdoor mini-kitchens popular in Florida. These generally include a built-in gas grill and sink that supplements the primary indoor cooking center of the house. An easy pass-through window between the kitchen and the deck is also a nice consideration for outdoor dining.

6. *Consider the benefits of covering some or all of the deck.* Retractable awnings offer an attractive, practical solution to covering a deck area inexpensively. When properly installed, they also add color and character to the appearance of a deck. Awnings can be implemented over at least a portion of the deck to provide some protected area, while leaving other sections of the deck open to the elements.

7. *Design decks and patios such for future use as a screened porch.* The immense popularity of screened porches that keep insects from the outdoor space should be considered. Can a roof or porch enclosure over the deck be added at a later date, along with a screen concept?

Outdoor spaces around new houses are too important to be ignored or considered as an afterthought in the design process. Affordable and modest houses need more of the attention typically given to larger homes; this does not necessarily mean more construction dollars, just a more balanced budget between indoor and outdoor spaces of the house. There are many readily available guides on popular deck and outdoor patio design that are inexpensive to implement.

Fences and Walls

One of the consequences of smaller lots and higher density homes is the increased need for *privacy walls* to define individual lots. Historically, privacy screens between houses were established effectively through landscaping: hedgerows, bushes, low-but-dense shrubs, trees, etc. Today, considering the cost or work involved in adding or maintaining these plantings, the alternative of a maintenance-free masonry or wood/vinyl fence, provided by the homebuilder, can seem preferable to buyers. Additionally, fences may be mandated by the market in higher priced homes or by the local government in lower priced homes.

A drive through a modest subdivision that has paid little to no attention to fence design or a control policy can be disappointing. Fences of different heights and appearance can give the community a hodge-podge look, particularly when they encroach on the front yard. Homebuilders cannot ignore the typical homeowner's intense interest to establish a territory or at the very least develop a measure of privacy in the rear of their property. People will invariably want to build fences around their yards, particularly in small-lot communities.

Fig. 4.33 (a) Historically, landscape elements such as hedgerows were planted to develop privacy between houses, but today (b) ugly fences seem to be more common

Fencing and walls can be supplied and even controlled in many different ways depending on regional customs and market categories. In the Southwest, for example, walls tend to be provided by the homebuilder and built of solid masonry, while in the East and Midwest, they are built after the sale by individual homeowners and commonly are made from wood. In either instance, the following suggestions apply for privacy fences and walls around production houses:

1. *Implement a fence control mechanism for the community.* If fences or walls are to be built individually by homeowners, establishing a control mechanism similar to architectural guidelines—enforced through a homeowners' association or deed restrictions—helps to address design issues such as height, materials, decorative motifs, and location.
2. *Incorporate some landscape material into fence designs to soften their impact.* Landscape material is expensive to install and requires ongoing maintenance, but, if used in limited quantities at key points it will help reduce the harsh lines of a solid fence. Small trees, low shrubs and vertical accent plantings are just a few examples of effective screen landscape materials.

Fig. 4.34 Privacy walls at *Providence Park* in Myrtle Beach, SC were integrated directly into the exterior walls of houses, with consistent materials and colors. Wall in front yards could be no higher than 3 ft, while side walls were 6 ft high. A brick cap, stucco finish, and plantings helped to soften the wall's appearance

3. *Use articulated fence designs.* If the homebuilder provides fences or walls, the design should have relief in the form of patterns, varied materials, or other design techniques that soften the visual impact of a long expanse of wall.

4. *Keep fences as low as possible.* Fences or walls located in the front yard area should be lower to the ground, on the order of 3–4 ft in height. Other fences should be no higher than necessary for privacy, generally no more than 6 ft.

5. *Encourage small stretches of fence to define lot corners, entry points, and other edges.* The idea of building a "hint" of a fence at key points to define boundaries is less expensive and less obtrusive on the rest of the community than a full length of fencing. Why install a fence around the whole perimeter when you can get the point across with small lengths that serve as boundary markers?

Higher density homes with privacy walls should incorporate the walls into the floor plans and lot layout for the initial concept stage since there are maintenance and control issues in addition to design considerations. One community where design was based on a privacy wall system integral to the architecture was *Providence Park at Antigua*, a higher density community tailored to empty nesters in Myrtle Beach, South Carolina.

At *Providence Park*, lots were 50 ft wide by 100 ft deep. The design program called for ranch plans varying from 1500 to 2200 sq. ft. In order to maintain privacy in the rear lot area, "courtyard" homes were designed with integral masonry garden walls that connected to the exterior wall of the home. Wall materials (stucco on concrete block) were strictly controlled, and wall heights were designated to be no more than was needed for privacy: 3 ft in the front and 6 ft in the rear and sides.

Accessory Structures

Another major planning and design issue that has become even more acute with small lots, tight garages, and houses without basements or attics is the need to accommodate outdoor *accessory structures*, or freestanding buildings on the lot. The most common examples are the ubiquitous "storage sheds" purchased at big box stores (and plopped in the rear yard of most suburban homes), jungle gyms, playhouses, etc.—some of which can be quite imposing. Other homeowners may opt for premade gazebos or other prefabricated structures.

There is almost a basic human instinct for homeowners to want to add additional buildings or structures around their personal residence. Whether functional or nonfunctional, secondary buildings can be beautifully designed to complement the main structure—as we saw with the historical family farm, which was really a collection of buildings instead of a single house. But most of today's homeowners can find neither the time, money nor expertise to build site-specific structures, so inexpensive prefabricated versions are more commonly the solution. Quite often these accessory structures can be eyesores for neighbors.

Fig. 4.35 Homebuilders should anticipate the natural desire that many homeowners have to add accessory structures on their lot, such as prefabricated storage sheds that can be unsightly for neighbors

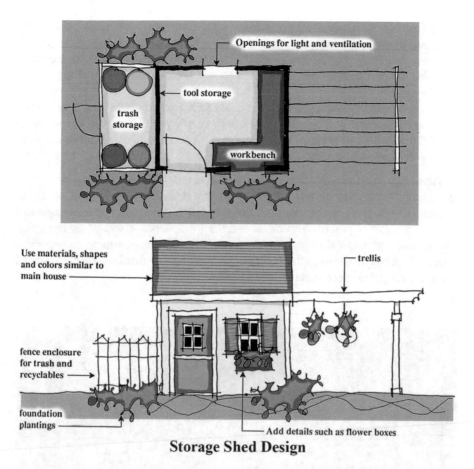

Storage Shed Design

Fig. 4.36 Sheds and other backyard structures can enhance and complement the main house if they are made using compatible materials and use design themes that add charm and human scale

Accessory structures are similar to fences: unless controlled by homebuilders, they have the potential to positively or negatively impact the character of the entire neighborhood. At the same time, their immense popularity indicates that accessory structures do need to be accommodated:

1. *Provide attractively designed prototypes as an option.* Independent structures that provide for storage, a workspace, play, etc. would be an attractive option to offer buyers of presold houses. Designs could then be coordinated with the materials and colors of the main house.
2. *Refer buyers to stores or craftsmen who stock or build well-designed structures.* For builders who don't want the task of providing optional accessory structures, refer buyers to those who can.

3. *Provide a prototype design package for accessory structures along with suggested fence and landscape concepts.* Ideas for fence designs and sheds are encouraged to be consistent throughout the neighborhood. These could be recommended by the builder and controlled by a homeowner association.

4. *The need for accessory structures could be reduced by providing ample storage space in a garage.* One option is to simply prohibit accessory structures through deed restrictions or homeowner association rules. In doing so, however, storage needs must be satisfied within the main structure of the house.

Accessory structures should be viewed as opportunities to introduce some delight into production houses by providing a secondary destination within the lot. Older houses with detached garages are good examples of this: the garage can be expanded into any number of ground-level or second floor functions—a workshop or garden area below, a granny flat or guest suite above. These types of structures provide a retreat-like atmosphere from the main house. New houses that include rear-located garages could offer these same opportunities.

Fig. 4.37 This rear yard playhouse by Glen + Williams Architects won a national AIA Honor Award. (Courtesy J. Lee Glen and Roxanne Williams, formerly Glenn + Williams Architects)

Other concepts for secondary structures may take the form of a playhouse or other simple structure scaled to a child's world; a private guest room suite that doubles as a home office; a workshop that may be combined with a garage; or a cabana for an outdoor pool or spa that may also include a small outdoor cooking area. As with fences, there are many books chock-full of design ideas for backyard structures.

Accessory structures are another example of an amenity that could be offered by builders on an option basis or ordered by purchasers in advance from a stock menu. Although some zoning regulations may not favor accessory structures, most will permit them with setbacks more lenient than those required for the main house.

Landscaping

Most production houses have little or no landscaping included in their price. Traditionally, landscaping was installed by homeowners over time to suit individual tastes and budget. Today, problems with landscaping are similar to previously discussed issues (such as fences and accessory structures)—homeowners need help, and homebuilders should get them off to a good start with a minimum landscape package.

This problem is more obvious in affordable housing markets. New owners, financially squeezed with their new mortgages, may not install any landscaping for years. They may be unfamiliar with how to establish a lawn, trees, or shrubs. This can result in overgrown weeds or soil erosion on a smattering of lots throughout a new community, bringing down its overall appearance.

Progressive homebuilders are now addressing the landscape issue in their products. More and more houses come with installed lawns. It is increasingly common for builders to include a landscape maintenance service agreement for the initial year of ownership as part of the purchase price. Trees and shrubs are becoming more mature at installation to avoid the "lone twig" sticking up in the front yard. Consider some of the basic landscaping suggestions that follow:

1. *Foundation plantings are important.* They make a graceful transition between the structure of the house and the ground plane. Larger shrubs may also help hide utility meters, compressors, and other unsightly elements that by law need to be located in front of the house.

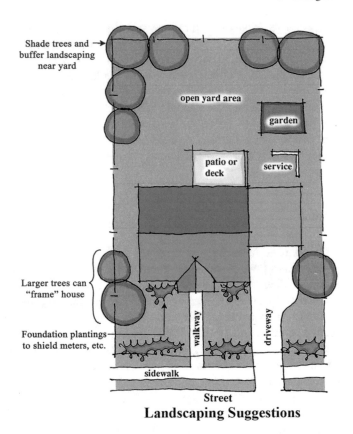

Shade trees and →
buffer landscaping
near yard

open yard area

garden

patio or
deck

service

Larger trees can
"frame" house

Foundation plantings
to shield meters, etc.

walkway

driveway

sidewalk

Street

Landscaping Suggestions

Fig. 4.38 Builders should help new homeowners get started with landscaping the area surrounding the house—it is particularly important in the more affordable housing markets

2. *Shrubs between closely spaced houses can add privacy.* Evergreens can help shield views or add privacy between houses, particularly along heavily used outdoor areas.
3. *Several larger trees and shrubs can help "frame" the overall house.* Saplings are commonly placed in front of the house—and later grow to a size where they can hide it! Consider how to locate plantings to enhance the street view of a house without any potential for future blockage.
4. *Coordinate landscape materials with street and sidewalk plantings, if any.* If street trees were not planted by the community developer, locate them along the road with reasonable spacing if possible.
5. *Rear yard landscaping can be left more to the owners.* It is still helpful to add several trees or shrubs here that will mature over time for the new owners, particularly if no natural trees remained on site during construction.

Basic landscaping, like fences and accessory structures, should be addressed in some fashion by production builders. Even in the most affordable new home community, homebuilders can help with written guidelines, suggested contractors, and small allowances. Unfortunately, landscaping is all too often left completely out of the building process, not in small part to the fact that is the last phase in the construction sequence—and budget.

Future Lot Patterns in Production Homes

As production housing moves to respond to human and social issues, we may expect siting and lot patterns to return to pre-war standards of smaller and more reasonably sized lots. I feel we will see more front–rear lot orientations and fewer narrow-lot concepts. The automobile will be accommodated more sensitively, and a more favorable balance will be achieved between community and privacy needs.

In the next chapter, I will begin to look inside the house at floor plans and room relationships that determine the physical appearance of the house on the lot, and its appearance from the street and neighborhood.

Note

1. Kendig, Lane. 1980. Performance Zoning. Washington, DC: Planners Press. p. 225–226.

Chapter 5
Floor Plans and Building Image

At this point, I would like to move inside the house to discuss programming room locations and how they relate to the overall house and lot. Since the interior room organization pattern generally establishes the exterior appearance of the house, the overall building image is integral to discussion here. The goal is to balance interior and exterior objectives through the design process.

If you look through all the house plan concepts you can find, you'll see that all plans conceptually fall into one of two categories: box-like or articulated. Box-like plans are organized as a collection of rectangular shapes covered with interlocking roofs and are typically used in affordable or modestly priced housing in the Northeast, Midwest, and other locations that have basement foundations. Articulated plans, on the other hand, are differentiated by numerous breaks in the foundation, walls, and rooflines. Articulated plans are free in their layout and are typically used for upper-priced and custom homes. They are more common in the South, Southwest, and West due to their shallow foundations, where building jogs are less expensive to build.

Today's box-like designs may be associated with the historic simple folk houses that were built throughout the country and served as affordable housing stock for most of the population. Articulated homes, on the other hand, were better represented by the stylistic homes built in various periods for the wealthier classes. In many ways, these still distinctions hold true today.

Certainly, there are plan types other than box-like or articulated—circular shapes such as igloos, tents, and geodesic domes. But these are anomalies—most production houses are based on plans that can be constructed with lightweight balloon framing—an industry standard for over 100 years—that favors one of these structural approaches.

For the most part, I will focus on design issues related to box-like plans, primarily because they still represent the majority of the modest and affordable production homes built today. Articulated plans typically give the designer more freedom to

Fig. 5.1 Most house plans are either (**a**) articulated or (**b**) box-like in character. Historically, box-like plans served more affordable housing, while articulated plans were varied in style and built for the wealthy

© Springer International Publishing AG 2017
J. Wentling, *Designing a Place Called Home*,
DOI 10.1007/978-3-319-47917-0_5

locate rooms and walls without being hindered as much by structural concerns, as they are more plastic in nature and thus give the experienced designer more freedom to develop a pleasing effect, elevation-wise. Many of the design concepts I will discuss, however, can be applied to either box-like or articulated plans.

Let's start our discussion with an extremely basic plan, one that might commonly be used for affordable modular housing due to its simplicity: a rectangle. If we assume that the appropriate rooms can fit inside the rectangle in a livable fashion, the threshold question becomes "How can we make such a basic plan look interesting in three dimensions? This may be affordable housing, but can't we do something to give the house more character and presence?"

Breaking Out of the Box

Even the simplest plan needs some relief from the tyranny of minimal structure and cost concerns. What it needs is to "break out of the box." Something must be added to the plan or elevation to upgrade it from utilitarian shelter to a more personable looking dwelling that feels like a home.

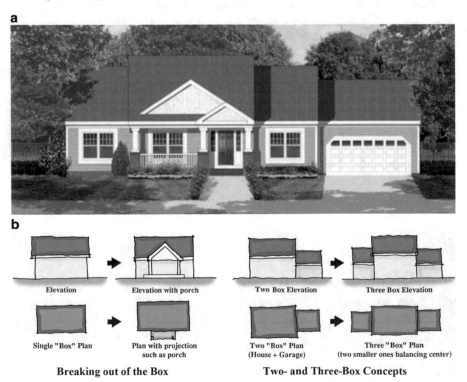

Fig. 5.2 Even the most basic plan needs some relief from the tyranny of minimal structure and cost concerns. (**a**) Here, the addition of an entry porch breaks up the longitudinal axis of this modest house for the *Brentwood* community in Bozeman, MT. (**b**) Houses can be viewed in the abstract as a collection of boxes that are connected to form an overall composition

In designs for affordable single-level homes, we start to address this issue by adding a projection on the front elevation that can break up the longitudinal axis of the roof. Typically, this is an entry porch of modest depth but of significant width to make a statement on the front façade.

Moving up the cost ladder from a purely exterior tack-on such as a porch, one can consider a projection in the interior plan that could allow a gable to be expressed on the front elevation. It's a good idea to make such a projection significant in size—each jog in the foundation and wall is costly, and when a projection occurs, it should be large enough in proportion to the balance of the façade to make enough of a visual impact to justify its cost.

A Collection of Boxes

The next more complex level of plan and elevation development is to consider a house as a collection of two or more boxes or rectangles that are connected to form an overall composition. The simplest example of this is to consider the main house as one box and the adjacent garage as another. Connected together, they form a composition that will be viewed as a whole. If you go to the next level of complexity, you might add a third element on the other side of the house to balance the garage. Now, you have a three box concept.

If you recall the discussion of *Mount Vernon* in Chap. 1, you'll remember it is composed of three modules. The larger main home is flanked by two smaller wings connected to one another by open arcades. This approach is found in some larger production homes that have a two-story central element balanced by one-story sections on either side, an image still seen as ideal.

The analogy of houses as a series of boxes or blocks may seem oversimplified, but it can be helpful to start a design process by considering the overall massing first, then work toward refinement of each mass. These principles can work for the smallest of affordable homes right up through larger stylistic mansions.

Structure and Light

In the early phases of organizing a house plan, one of the governing considerations in sizing boxes and rectangular shapes is *structural limitations*. To resolve structural limitations, go right to the top of the house and look down at the roof. Most plan shapes are limited in dimension by reasonable roof framing spans, typically determined by the spans running parallel to the roof slope. The roof is generally framed with either roof trusses or roof rafters, both of which might limit the overall span to, say 40 ft. Most homes are designed to have roof spans ranging from 24 ft for affordable housing to 36 ft or more for larger homes. Floor framing inside the rectangle is a secondary consideration in resolving structure because floors can be supported with intermediate bearing lines provided by beams or interior walls.

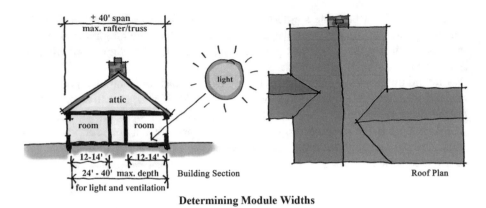

Determining Module Widths

Fig. 5.3 Structural and light penetration issues govern basic sizing parameters of floor plan modules

Roof framing spans generally coordinate with the other major plan consideration: *bringing natural light into the home*. Houses, unlike office buildings, stores or factories need to have virtually all major rooms on an outside wall with access to light and ventilation. In fact, the more rooms that have plentiful natural light, the more livable the home becomes. Therefore, plans that are based on rectangular structural shapes are also governed by dimensions that will allow light to penetrate into rooms. Assuming a rectangular plan can contain a room in the front, a room in the back, and a hall in between, the desired depth of the rectangle should not exceed twice the widest dimension of typical rooms plus a hallway. This again translates to about 2×12–16 ft, or 24–36 ft as a maximum module dimension.

Establishing the ideal footprint for a house is a critical step in the design process. What should be the width of the house in relation to the lot? How about depth as governed by structure, light, lot size, and program? How articulated can the shape or shapes become and still meet the cost constraints for the selling price? Where should key rooms be located? These are programming questions we will attempt to answer in the balance of this chapter.

Establishing Primary Rhythm

Let's assume for cost purposes that we need to stay within one basic rectangular plan, and that we can afford to build only one projection along the front elevation. Let's also assume we want the same plan with several different alternate elevations available. What can we do to design an attractive home?

The initial step would be to look at establishing a major "rhythm" or "modulation" for the plan. This is typically accomplished by organizing building shapes, projections, windows, and other design elements into a pattern that has some *order* to it—instead of just a haphazard composition. Looking at our simple ranch home in Fig. 5.4, if we put a gable on the front wall, we immediately establish at least two *modules*; gable and wall. If we put the gable in the middle, we now have three modules; wall-gable-wall. We could put a gable on one side, a porch in the middle, and nothing on the other side; then, you have gable-porch-wall.

The amount of modulation in the massing of a home will vary with the size and complexity of the program. Smaller and affordable houses may only have two or three modules, while large mansions may have four to six or more. This is a function of design philosophy; generally, the more modulation in a façade, the more it can relate to human scale. Tudor style homes, for example, tend to be very large, yet highly modulated with different forms and rooflines. Georgian style homes tend to be more massive and unbroken in scale. These are simply different philosophies of rhythm in design.

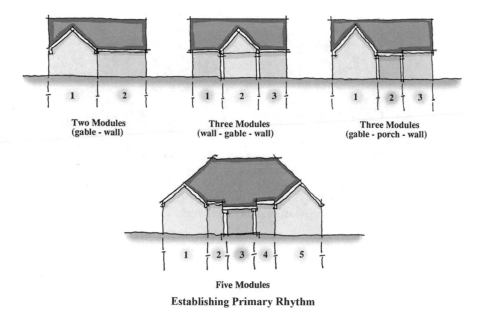

Two Modules
(gable - wall)

Three Modules
(wall - gable - wall)

Three Modules
(gable - porch - wall)

Five Modules
Establishing Primary Rhythm

Fig. 5.4 An initial step in floor plan development is establishing a rhythm for the exterior of the house

Symmetrical vs. Asymmetrical

As mentioned in Chap. 1, *symmetry* is one of the most time-honored organizing principles for houses and architecture. Symmetry can be used to incorporate a sense of order and balance into an otherwise utilitarian plan. By the same token, *asymmetrical orders* can establish an equal sense of design purpose and possibly allow more freedom in the plan's layout.

Back to our basic rectangular plan. If we locate the porch directly in the center of the home, leaving equal space on either side, we now have a primary order of symmetry. By locating the porch to one side, we have a deliberately asymmetrical plan and elevation effect from the front. Both orders, however, are based on the same plan.

Designers can use the juxtaposition of symmetrical and asymmetrical elevations to make the same model house look different. When considering that a new community that will feature three or four model homes, the combination of symmetrical and asymmetrical alternate elevations will help to enliven the street with varied façades.

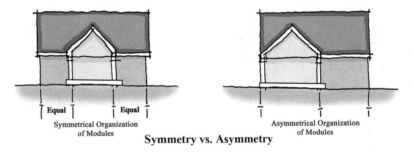

Symmetry vs. Asymmetry

Fig. 5.5 The concept of symmetry or asymmetry can be effectively used for different reasons: symmetry suits formality and simplicity, while asymmetry suits more formal and elaborate styles

Secondary Rhythm and Orders

Continuing with our simple rectangular plan, let's look further at the symmetrical plan order of wall-gable-wall. We now have three modules—a central gable and a wall on either side. Now look at each of those modules in terms of *rhythm* and *order*, primarily in *window placement*. Assuming you have established symmetry for the overall house with the three major modules, you would want to reinforce that theme in each individual module. Therefore, if you put one window in the wall on the left side of the gable, you would want to balance that with the same window on the right side of the gable. Or if you put two on the right, you would want two on the left, and so on.

The windows within each module can and should together create a *secondary order that reinforces the primary order*. If you have two windows, you want them to form a composition by either being placed together or spaced apart equally within the module's wall space.

These same secondary orders can be established with an asymmetrical plan that may have only two modules. Within each, you would establish an order that reinforces that module, but need not match the other. For example, let's say you put one central window on a wall module and a door and a double window in a gable module—those secondary elements should reinforce the primary module.

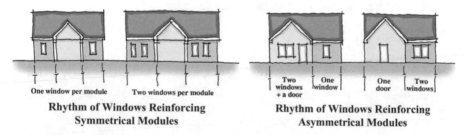

Rhythm of Windows Reinforcing
Symmetrical Modules

Rhythm of Windows Reinforcing
Asymmetrical Modules

Fig. 5.6 Window and door placements, along with other elements, can help to establish a secondary rhythm or order within the primary order. In these two cases, window placements reinforce the primary order of the plan modules

Vertical and Horizontal Orders

So far we have been concerned only with rhythm in the horizontal dimension. Introducing the next level of complexity, let's expand our rectangular ranch plan into a two-story home and talk about *integrating vertical and horizontal rhythm and orders*. In its simplest form, the second story can repeat the rhythm of the first. For example, we could take our symmetrical plan and just add another floor on top and repeat the window spacing and pattern on the second level. The entry door location on the second floor could be omitted, or become a window of some sort that need not match the others, or could be a door to a balcony over the entry door.

Let's say the second floor cantilevered over the first floor by 12 in. Or that there is a roof between the first and second floors. Now, you start to define the second floor as a separate order. Perhaps the first floor has asymmetrical elements and the second floor is symmetrical—or the first floor is brick and the second floor is stucco. The complexity of potential orders is now double that of a one-story home.

Another variation to the process of establishing vertical and horizontal orders occurs when combining one- and two-story elements within a plan—so the façade may be dominated by *either* vertical *or* horizontal orders. Let's say you connect a one-story element next to a two-story component. You can reinforce the verticality of the home by making the two-story component symmetrical vertically within its own module. For a horizontal look, you can establish an order within the one-story module, then treat the second level as a separate order.

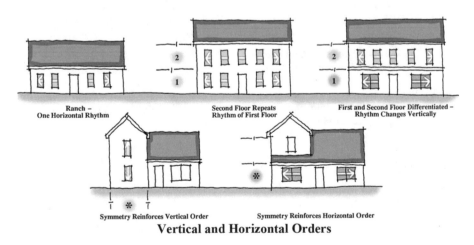

Ranch –
One Horizontal Rhythm

Second Floor Repeats
Rhythm of First Floor

First and Second Floor Differentiated –
Rhythm Changes Vertically

Symmetry Reinforces Vertical Order

Symmetry Reinforces Horizontal Order

Vertical and Horizontal Orders

Fig. 5.7 Horizontal and vertical orders should be reinforced in a larger house façade

Fig. 5.8 This modest house in Durham, NC is organized as three modules, with each module's rooflines, window and door placements further reinforcing the façade's organization

Complexity in Production Housing

In general, the larger and more expensive a house, the more complexity one might find in the orders and rhythms. This is why historically articulated styles such as Tudor and the French Country styles are still popular in the upper price ranges. Simplicity, on the other hand, can be a good design ally of the smaller and lower priced houses, as seen in early folk houses.

Therefore, one objective in plan organization is to provide the appropriate level of complexity to establish a perceived value for the house. Affordable homes may rely on less complex floor plans reinforced by symmetry, while more expensive homes may have many levels of complexity to develop more articulation.

Complexity in the building elevations can include both large- and smaller scale elements. In this chapter, we are discussing the larger massing issues such as module size, roof height, window placement, and so on. In Chap. 7, we will address details that can be incorporated into the elevations that will further enhance the level of building complexity.

Front Elevations vs. Side and Rear Elevations

Up until now, we have been discussing issues of rhythm, order, and balance by primarily considering the front façade. Unfortunately, this is generally the only elevation where these issues are addressed. Next time you drive along a suburban street, take a look at the side and rear walls of houses. Chances are you will see no sense of order in window placement or other elements.

Fig. 5.9 (**a**) Larger houses tend to be more complex in plan shape, while simplicity is an ally of more affordable houses (**b** and **c**)

Most residential designers do not dwell on side and rear elevations. This ethic is in sharp contrast to historical designs where most homes had some order on all elevations—not just the front. If you take a drive down a street in a prewar subdivision, chances are the designs have better organized side and rear elevations.

The reason for this problem was discussed earlier—the priorities of new home design do not value the exterior appearance of the home as much as the interior. Plans are formulated based on interior layouts with windows placed in a random pattern to suit the interior plan. Only in higher priced homes, such as those found in golf course communities where side or rear elevations are quite visible, are elevations other than the front considered.

Ideally, all sides of a house should have the same sense of rhythm, order, and balance found on the front façade although perhaps not with the same grandeur. Side and rear elevations may include elements such window bays, decks, and other projections; however, these should also work within a composition to form a pleasing elevation. This is a particularly challenging design task with a narrow lot and other higher density homes that incorporate blank walls and other eccentricities.

Fig. 5.10 (a) Historical designs generally had a sense of order on all elevations, as seen on this side view of an older house. (b) Today, random placement of windows or other elements on side elevations is common

Inexpensive ways to introduce order into side and rear elevations include the following:

1. *Where possible, line up first and second floor windows.* On two-story façades, try to line up first and second floor windows, even if some interior adjustments need to be made. This helps make the home appear as if there is a balance between the inside and outside of the house.
2. *Treatments of the side and rear windows should match window treatments on the front façade.* One thing I find really irritating is the transition from highly articulated windows on the front façade to totally undecorated windows on the side and rear of a home. If the front windows have muttons and trim, then the side and rear window treatments should match. The additional cost here is negligible.
3. *Locate windows, doors, and other elements on side and rear elevations in an ordered pattern.* For example, if you are placing a window on the side wall of a garage, consider locating it in line with the ridge of the roof. When locating a projection on an exterior wall, consider its location relative to rooflines, windows, and other elements to create an organized appearance.
4. *Consider using gable vents or gable windows on the side and rear elevations.* These help to add some architectural decoration to what can become a large expanse of wall area. Gable vents are becoming less common with the advent of the more functional roof vents; however, the decorative value of adding a distinctive gable vent should not be overlooked.

Façadisms

Major problems are apparent in many new home designs when too much or too little "design" is ineptly applied to the outside of the home—primarily on the front façade. With the view that the interior plan reigns supreme, exterior design is applied after the fact, if at all.

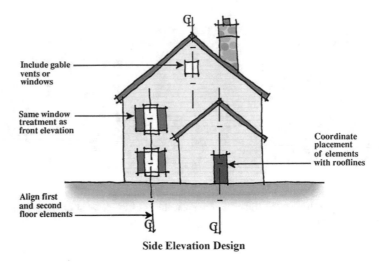

Side Elevation Design

Fig. 5.11 Side and rear elevations need design attention at a level consistent with the front façade

Some of the most frequent elevation goofs are as follows:

1. *The Bauhaus Look.* Little or no attention is paid to the exterior design of the home. Walls, doors, and windows are placed in a random pattern and do not follow any order. The typical excuse used to justify such a design is "hey, these are affordable homes."

2. *The Overdecorated Look.* This is the opposite of the Bauhaus Look. Goal here is to articulate the front façade with many gables, materials, window sizes, and other ornaments to create a sense of "value." Instead, the house looks overdone and confused.

Fig. 5.12 Some common blunders in elevation design, such as (**a**) the "Bauhaus Look" and (**b**) the "Over-decorated Look," make for a poor streetscape presence

3. *The Cliché Look.* Here, the designer incorporates the latest in architectural cliches that add no value to the home, yet may in fact date it to a particular era. These are most apparent in new home designs as a designer's "signature" or "trademark."

4. *The Close-but-no-Cigar Look.* Here, the design is almost accomplishing something—but stops short. For example, a façade that looks like it should be symmetrical, but just didn't quite make it. The problem is most likely that the builder or designer didn't follow through.

5. *The Wrong Style Look.* Here, the designer may try to recall a particular historical style but incorrectly interprets the style by missing the original intent, wrong proportional relationships, mixing elements, using inappropriate materials, and other missed shots.

I could go on and on with this, but you get the idea.

Standard Plan Arrangements

It has been argued that if you look at all the house plans in the world, that there are really only four or five basic variations out there. There are myriad minute differences within those plans, but, conceptually, they all can be traced to a handful of stock layouts builders have favored for years. After all, you only have so many rooms—how many different ways can they be arranged?

In my earlier book, Housing by Lifestyle, The Component Method of Residential Design, I suggested that residential floor plans can be organized into five basic *zones,* or *components*: Ceremonial, Community, Privacy, Functional, and Outdoor areas. Each of these zones contains a series of rooms or areas that varies according to market considerations. The method of organizing these components is governed primarily by the building or lot pattern, and that then determines which orientations will be public, which will be private, which rooms will have views, prime natural light, and so on.

These organizing principles are relatively simple and are based on living preferences that have been validated by centuries—ceremonial rooms such as dining rooms are better located in the public sector of the lot, private rooms such as bedrooms should be located in the interior or rear portion of the lot, and functional areas such as garages, laundry, and storage should be in accessible, but not prime locations.

Therefore, within regional markets where lot sizes, square footages, and prices are relatively the same, common sense layouts tend to work their way into four or five basic patterns. In many locales, the same basic three or four plans have been built by the hundreds of thousands for decades. Today, new smaller lot patterns, affordability concerns, and changing household compositions are challenging basic plans. In theory, house plans are constantly evolving to reflect mass-market values, tastes, and lifestyles.

One major plan shift, for example, is the increasingly popular *first floor in-law suite* in a two-story home. This plan, which was somewhat rare in most markets a decade ago, is now quite popular in markets with households who want to add space

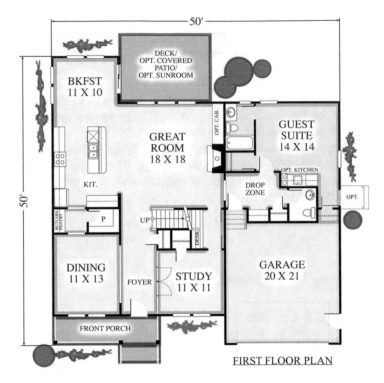

Fig. 5.13 Although most markets are dominated by four or five standard plan arrangements, new layouts such as the in-law suite plan are working their way into even the more design-conservative locales

for an aging parent with a degree separation of separation between bedroom areas. In the Sunbelt, the *pavilion plan*—where smaller rooms or *casitas* are physically separated from the rest of the house to serve as a more private guest suite or office— is another recent innovation that is driven by new lifestyle preferences.

Indoor/Outdoor Relationships

Interior room locations should actually be governed by whatever is outside the room. The entry hall generally needs to be in the "front" of the lot. The family or living room should be near the main yard or patio area. The kitchen should also be near the main yard or patio, as well as the garage or parking, in car-dependent communities. *Indoor/outdoor relationships* are primary organizing principles for livable plans, due to the need to have the best *natural light* and *outdoor access* adjacent to the rooms that are used the most.

It's a shame so many homes relegate some of the most lived-in rooms to an interior or otherwise poor location without good access to light, views, or ventilation. In most plans, the community component, which includes the kitchen, breakfast

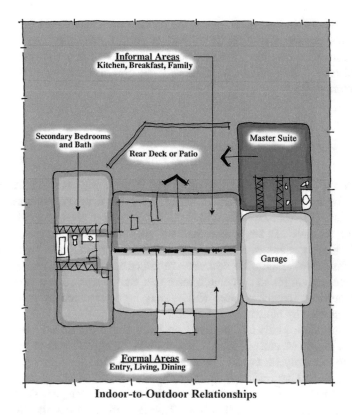

Indoor-to-Outdoor Relationships

Fig. 5.14 Interior room locations should be governed by whatever is *outside* the room. The most lived-in rooms should get prime exterior orientations relative to the rest of the house

and family/great room, should get the best orientation, with plenty of windows to view the primary yard or outdoor areas. For most floor plans, the common practice of combining the kitchen, breakfast and family/great room into one large component allows more interior space to share in the best outdoor orientation.

Infrequently used rooms do not need prime locations. For example, in more and more plans, the formal dining area (which may only be used once or twice a year but still is considered important to buyers) is one of the spaces that becomes internalized or given less than prime orientation. This is perfectly normal in floor plans that need to prioritize room placements. Other rooms that may be given poor outdoor orientations include secondary bedrooms and flex rooms.

Some spaces need outdoor access more for functional purposes instead of just for light and views. Utility rooms work best if they have direct outdoor access to a service area. Garages benefit from having direct access to the outside through swing doors instead of just large pull-up doors. With the trend toward working from home, a *home office* will benefit from having a direct door to the outside, ideally near the formal entry, is often desirable for visitors.

Particularly in Sunbelt ranch plans, the more rooms with direct orientation and access to the outside, the better. If the budget permits, virtually every room could access a courtyard or patio. If the homes are designed for the affordable market, additional access doors could be offered as an option or shown as a later home improvement project.

Narrow Modules and Spans

In order to maximize exterior orientations and the penetration of light into interior rooms, there are two rules to remember: (1) keep the depth of basic modules to a minimum and (2) locate key rooms to have two exterior walls that can then be punctuated with windows. In larger and more complex plans, designing a plan as a collection of narrow modules allows the exterior wall to undulate, maximizing the potential for access to light. These narrow modules help the elevations impact on the home as well, giving it an articulated form that can be treated in a variety of ways from the outside. Ideally, module widths may be determined by just one or two rooms, connected by hallways.

Narrow module plans can be used on even the most affordable homes. We designed a series of entry-level bungalow style "shotgun" houses with modules which were primarily 14 ft wide. These houses, built at the *Breckenridge* community in Durham, North Carolina, included plans that were a collection of narrow modules connected together in one and two-story configurations to both provide interiors with generous natural light. Although only 14 ft wide from the street, the closely spaced narrow façades gave the houses a collectively strong presence from the street.

The sideyard homes of Charleston, South Carolina provide another example of how narrow module houses can be more livable than conventional designs. Charleston houses are typically only 16–24 ft in width with porches along one side of the house on both floors. This is really the same idea behind contemporary zero-lot-line houses, but in the narrow Charleston houses all rooms face a side court. Replicas of Charleston houses are still built today and have been enjoying a revival in traditionalist communities such as seen earlier at *Harbor Town* in Memphis, Tennessee.

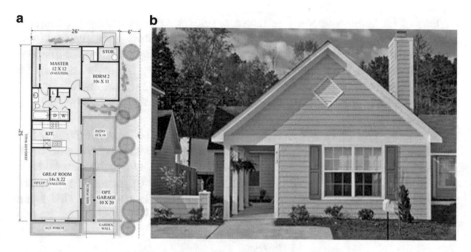

Fig. 5.15 At *Breckenridge*, these very affordable houses were primarily 14 ft wide, reminiscent of Southern "shotgun" houses. (**a**) The narrow spans helped to maximize penetration of light into rooms. (**b**) Such a plan allows more rooms to view the private outdoor courtyard in the sideyard area

Fig. 5.16 Replicas of Charleston sideyard houses were still being built in the 1990s in communities such as *Harbor Town* in Memphis, TN. Incidentally, this Charleston sideyard was the Best in American Living National Design Competition's House of the Year in 1991. (Courtesy Looney Ricks Kiss. Photo by Jeffrey Jacobs, Mims. Studio)

One of the most common folk house designs seen in farmhouses throughout America is the simple two-story narrow plan with a one-story wing attached to the back. The two-story part of the house has the primary living rooms on the first floor and bedrooms above. The one-story wing in the back is typically the kitchen. The façade is symmetrical with a front porch and often a central gable with a window into the attic. (This house is in the background in Grant Wood's famous painting, "American Gothic.")

The farmhouse plan, with its narrow spans and well-lit room arrangements, provided a model concept for livability for new home designs at *Eirean Mohr*, an affordable community in Salisbury, Maryland. At *Eirean Mohr*, the historic farmhouse design was expanded to include a bedroom downstairs and an expanded kitchen/breakfast/family room in the one-story section in the back. This layout allows for much of the house to be able to access a generous deck along the one-story wing.

Fig. 5.17 (**a**) The narrow spans of the *American Gothic* farmhouse were an inspiration for this model house at *Eirean Mohr* in Salisbury, MD. (**b**) The floor plan had modules that were only 14'–16' wide and had a two-story element facing the street with a one-story element in back. This is the very same practical-yet-livable floor plan of many historic farmhouses

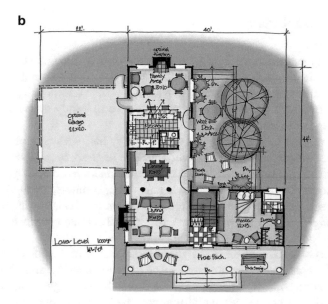

Fig. 5.17 (continued)

A Western example of the narrow module concept is the Eichler-built atrium home discussed in Chap. 4. The courtyard-oriented plans were typically one room in width (14–16 ft), with ample windows on two or three sides of each room, many of which were sliding glass doors accessing private courtyards.

Many new small lot configurations are forcing designers to configure houses into shapes that preclude narrow modules, but instead require a large plan shape that results in much of space as internalized and remote from natural light. These plan challenges can be addressed with concepts such as interior atria, and courts, similar to the Eichler houses.

Open vs. Closed Plans

One method to capture more light into interior rooms is to combine one or more rooms into a single space. This corresponds with the postwar trend toward floor plans that are more open and less rigidly partitioned than traditionally defined layouts. The open plan concept is particularly effective to jointly define some of the component areas discussed earlier.

For example, the formal living, dining, and entry hall—which typically form a ceremonial area—can easily be combined into one large space, allowing all rooms to share in the light and views. Floor plans can be made more open yet still have "defined" rooms with designs that include half walls, columns, archways, changes in floor surface, ceiling heights, etc. In conservative markets, buyers continue to prefer a strong definition between rooms, in most others, the open plan concept is welcomed as the desired approach.

Open plans make a lot of sense in light of today's view of the home. Many traditional designs continue to devote large spaces to formal living and dining rooms which are seldom used. By combining these rooms into a single area, they can be individually diminished in size but still appear large as a whole space within an open plan. For affordable housing, open plans make even more sense still—the casual lifestyle of most households favors open plans so families can easily gather and communicate.

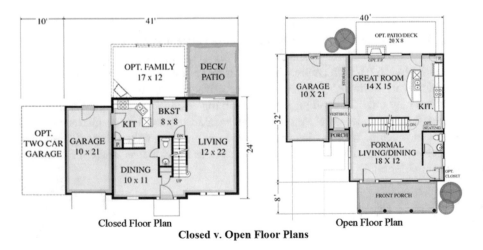

Fig. 5.18 Open plans can be used to borrow light and views from other rooms and respond to the preference for more casual living arrangements. One caveat regarding open plans is the issue of *noise*

We also see open plans on the second floor or where several bedrooms may be directly off a shared work/study alcove. Even laundry and utility rooms are becoming more open, and when combined with other work areas, they may connect to the rest of the house. After all, why should doing laundry be a task confined to a dark, cramped space? Why not have the ability to communicate while doing laundry or a working on a project?

One important issue with open plans is *noise*. The open laundry area is fine until you start running the washer or dryer. In that case, a solid door or double doors should be included. Pocket doors that recede into the wall are particularly appropriate for this situation. Often less considered, however, is the noise problem when the entire house is open—cooking, media, conversations, and other noises pervade the whole house. Here, closable partitions or screens should be included.

Circulation as Organizer

In most houses, a primary organizing system is the *circulation route* between rooms. This pattern is evident upon entering at the front door, where one is most likely to find the stair (in two-story homes) and/or a hallway to adjacent rooms. Similarly, there is a noticeably less formal circulation pattern throughout the house connecting the garage, kitchen, bedrooms, and other spaces.

Circulation plays a major role in establishing the character of any floor plan. For formal, more expensive homes, circulation spaces are usually well defined. Affordable houses typically integrate circulation space into the rooms in lieu of

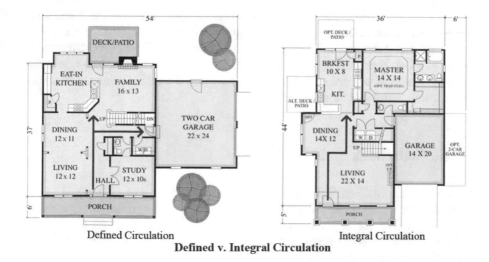

Defined Circulation   Integral Circulation
Defined v. Integral Circulation

Fig. 5.19 Circulation is another organizing system for floor plans. Circulation is more *defined* in larger formal houses than in smaller houses that can use living spaces for circulation as well

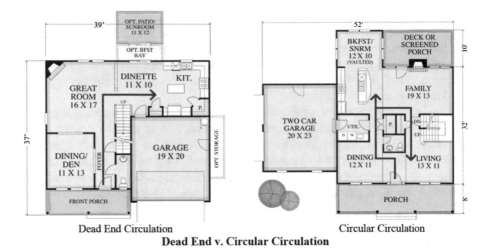

Dead End Circulation   Circular Circulation
Dead End v. Circular Circulation

Fig. 5.20 Houses with *circular* circulation, which offers two routes to get through the house, are generally more popular than plans with *dead-end* circulation

using valuable square footage for solely that purpose. In small houses, it's a shame to see precious square footage devoted to formal circulation at the expense of cramped interior rooms. It is also a shame to see houses where circulation paths destroy the usefulness of rooms by cutting directly through them. Many plans fail to accommodate furniture and circulation properly.

I have found over the years that people have a strong preference for houses that have *circular circulation*, meaning that within a plan there are always two routes to get to another room—as opposed to "dead-end" plans where there is only one way to move through a house. This preference is perhaps the reason why the "center-hall plan," with its multiplicity of circulation routes, has been so popular for centuries.

Circular circulation can be designed into a plan around the main stair in two-story plans, so that you can reach the stair from two alternate directions. In ranch plans, circulation is based on outside doors, allowing the paths to them to be accessible from several directions or rooms.

Front Stair/Back Stair

Just as a house can benefit from having a good location for a front door and a back door, the same can be said for having two stairways, even on homes of modest size. Two-stair houses can be seen extensively in historical plans where the kitchen was removed from the front of the house—and in large houses where second stairs were needed to accommodate servants.

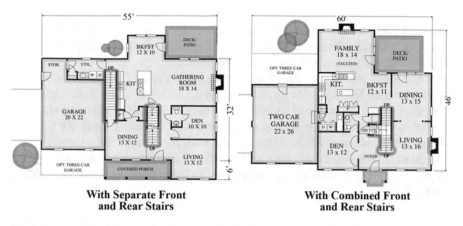

**With Separate Front
and Rear Stairs**

**With Combined Front
and Rear Stairs**

Fig. 5.21 The historical concept of including both a front and a rear stair in floor plans is returning to favor in even modest houses. The back stair can be completely separate, or it may join the front stair at a common landing

Front stairs and back stairs function in a complementary fashion to give houses an additional sense of character. The entry hall and *grand stair* is a very ceremonial space. Historically, this was an optimal location for a window seat, with lots of wall space and possibly shelves for display. The entry stair was a place to take the family portrait, enjoy greetings, and say long goodbyes to friends. The grand stair might include a mid-point landing, sometimes large enough for a table, perhaps a chair, or a second window seat. At the top of the stair, there may be a landing that overlooks the entry hall and adjacent rooms and may also have room for seating, tables, or shelves.

The grand stair may be built as a "T" stair, where a landing leads to two stair routes to the second floor, or two sets of steps lead to a common landing for a single set of steps to the second floor. In smaller houses, the grand stair may open up to adjacent rooms to become part of a larger space and be a stage-like place within that room.

The *back stair*, by contrast, is utilitarian. It is a direct route from the lower informal part of the house to the upper level. The back stair can be small and straightforward, or with minimal winding treads to save space. In some houses, the back stair may be a spiral stair. It has a totally different character from the front stair in both design and location.

Back stairs are regaining popularity in production homes that are as small as 2000 sq. ft. One historical technique to reduce the added square footage that a second stair consumes is to have the back stair link up with the front stair at a landing point (similar to a T stair) that then continues to an upper hallway. This solution positions the stair so that the base on the first floor from one side is the formal entry, and from the other side is the informal area. In larger homes, the back stair can be completely separate from the front and usually provides a more direct route from the kitchen/family areas to the secondary bedrooms.

Some current designs are favoring a single stair in a more informal location, such as the rear of the home, in a family room or hall off the kitchen. This design approach assumes that the stairs should be in the most used location for reduced traffic flow and more efficient circulation. If people spend most of their time in the informal section of the home, why should they have to walk through the formal area to get to the stairs? This solution can work well in casual and affordable markets, but the formal or grand stair will continue to be popular in the more traditional and upper-priced designs.

Floor Plans to Accommodate Lifestyles—And Furniture

Another organizing principle for floor plan development is simple to consider: how will people live in the house? Where will they spend the most time? Where will they eat? How will they eat—together or alone, formally or on the run? Where do they want to feel privacy or semi-privacy? How can a well-organized plan accomplish both the practical and emotional needs of a household with prototypical designs?

It is always amazing to observe how different people respond to different floor plans. Often a new home community will offer four or five plans, but inevitably one plan quickly becomes the most popular. In fact, it is more amazing how often this occurs *before any of the models are built*, just based on what people can sense from viewing the plans on paper. On brochures or in model homes, one often gets the feeling that some plans really work while others really don't.

What makes one plan work and another not work is the *degree to which design has successfully anticipated how most people want to live* in the plan. Where will

Fig. 5.22 One of the best ways to test the functionality of a plan is to furnish the rooms on paper. Curiously, this exercise is often overlooked in the design process

they want to relax, eat, entertain, and enjoy themselves in their new house? Where will they want high ceilings and maximum natural light? Where will they want work rooms and storage space? Where will they want fireplaces, if any? These preferences are clearly reflected in which models buyers purchase.

One of the ways to test the functionality of a plan is to "furnish" the rooms on paper. This exercise attempts to anticipate how people would arrange typical furniture pieces in a room or series of rooms. The inability to understand how a room will accommodate basic furniture is the downfall of many plans. Often interior designers work with preliminary architectural floor plans early in the design process to anticipate what modifications will be beneficial for optimal furniture layouts. But more often, furniture is not anticipated.

Some of the basic furniture items that need to be accommodated in a good floor plan:

1. *A living or family room needs to accommodate a sectional sofa or conventional sofa with chairs that form a conversation space*—with room for end tables and a coffee table around the chairs. The proposed furniture arrangement ideally will also allow a view of the outdoors, a fireplace, or television/

media screen (if any). Furniture arrangements work best if they allow communication with people in adjacent rooms of the house.

2. *A dining room needs space for a 6 ft by 3 ft (minimum) table with room for seating and circulation around the perimeter.* The dining room also needs lengths of solid wall around it to accommodate a serving table and possibly also a "china cabinet."

3. *A master bedroom or owner's suite needs to accommodate a king or queen size bed with nightstands on either side, along with two dressers and one or two chairs.* Those are minimum items of furniture. Space for a desk and table, and a defined "sitting area" with two chairs and a table, are desirable additions that can be accommodated in larger houses.

4. *A secondary bedroom needs space for a double bed, dresser, and desk with chair.* Again, these items are only the basics—additional space is always desirable.

Furniture pieces need to be equally anticipated in both affordable floor plans and large custom homes. Curiously, however, furniture placement is all too often totally overlooked in developing new plans. Furniture layouts help anticipate how people want to live in two dimensions. Another consideration is how a home functions in three dimensions. In other words, how should plans work to have communication between levels? A typical example is a loft overlooking a room below, or a two-story foyer or a family room. The third dimension of height can introduce the element of drama associated with changes in height.

Accessible and Adaptable Floor Plans

With the increased sensitivity to accommodating the needs of people with disabilities, the *Fair Housing Act of 1991* was passed, mandating that some classes of attached housing be made adaptable to the needs of disabled persons. As a result, the idea of *universal design*, or design standards that allow fixtures be used by

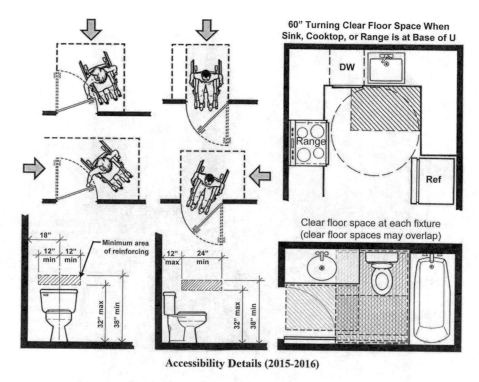

Accessibility Details (2015-2016)

Fig. 5.23 Making floor plans accessible for homeowners with disabilities is increasingly common. The movement toward *universal design* advocates floor plans to accommodate wheelchair access[1]

owners that want to "age in place," rather than move once mobility issues become a challenge for daily activities.

Universal design calls for slightly larger bathroom and kitchen arrangements with clearances for wheelchairs. Steps and grade changes are limited. Buyers then have the flexibility of installing lower counter tops, grab bars, and other aids that allow for independent living at their expense. So long as the primary design issues of room size, door openings, hallways, etc. are addressed to allow for adaptability within the finished construction, universal design can be easily achieved.

Small Rooms vs. Large Rooms

One of the big questions in programming houses concerns appropriate room size. It is just as simple to make a room too large as it is to make it too small. A certain size is optimal for the function of each room—a size that usually relates to the furniture arrangement. Beyond furniture accommodation, there is the question of the "emotional" qualities of various room sizes: which rooms should be large, and have

Fig. 5.24 This larger upscale plan has rooms scaled for the use of each space. Rooms are further defined by soffits, ceilings, and wall openings that can be finished with trim moldings. Note the internalized garage

higher ceilings, and which should have smaller, more intimate spaces. Frank Lloyd Wright made his entry foyers small, with dramatically reduced head height, so people would feel a burst of emotion when they entered the main living portion of the house.

Today, most consumer surveys indicate a preference for higher ceilings in some areas; the entry foyer area is often favored, followed by the great room, dining, and owner's suite. The extra height of vaulted or tray ceilings is sought to give these rooms a sense of character and importance.

Other rooms actually may work better small. Some rooms may need to be small to be intimate. The office/study is frequently thought of as being a cozy retreat and may actually benefit from smaller dimensions. Children's rooms, some of which are used for hobbies, are also often smaller in size. I often feel the typical home would be more livable with many smaller secondary rooms (such as bedrooms, flex rooms, and libraries) than few of generous size. I offer the following suggestions for specific dimensions for primary rooms in an average-sized house:

1. *Living/Formal Room.* 12 ft by 12 ft is minimum; 14 ft is more comfortable.
2. *Dining Room.* 11 ft by 12 ft—this space works well when combined with others, but still needs this minimum area. Greater length is better than increased width since the table length is more often varied than width.
3. *Breakfast Area/Nook.* 9 ft by 11 ft—this space can be smaller if using built-in seating or tables, in which case a 6 ft by 6 ft space functions quite well.
4. *Family Room/Great Room.* 15 ft by 16 ft; many are much larger than this, but often don't need to be in order to be functional, so long as a conversational seating area is provided.
5. *Home Office.* 10 ft by 12 ft. This could be an intimate space and still work at 9 ft by 9 ft.
6. *Master Bedroom/Owner's Suite.* 15 ft by 16 ft.
7. *Secondary Bedrooms.* 11 ft by 11 ft. This is generally considered a minimum, though I believe some could be smaller, perhaps 10 ft by 10 ft.

Room sizes should correspond to specific market needs; but, I continue to believe many rooms in a typical house are too big, and that smaller sizes may actually benefit the design's emotional qualities.

Lofts, Alcoves, Closets, and Secret Rooms

Another category of rooms that falls somewhere between small and large is *half-rooms* such as lofts, alcoves, "nooks and crannies" all of which are in effect subsets of a larger space. These half-rooms are very workable as semiprivate spaces where people can do things, yet still participate in the ongoing activities in the larger room. The best example might be a place for doing homework within earshot of the rest of the household. These smaller subspaces can add livability to a home by allowing for multiple individual tasks within a connected space.

Often *half-rooms* can be added to a design without increasing square footage significantly. For example, a second floor hall may need to be three feet wide for circulation purposes only. If you would add another two feet to the width of the hall, there would also be enough space for a small desk and telephone. At the top of a stair you might be able to add two to three feet and create a loft overlook with some bookcases to suggest a mini-library.

Small half-rooms might only be the size of a closet. If a room is planned with two closets, one could be converted to a small workspace within the room or a small hobby room. Children have been known to convert small spaces in closets into secret rooms for their imaginative play. Spaces such as these could be programmed into moderate-sized homes. Homebuilders should look for "hidden spaces" under roof framing before drywall is installed in their initial models.

Plan with Loft and Alcoves

Fig. 5.25 Half-rooms such as lofts and alcoves function as semiprivate spaces that, when used, are still connected to other places in house. A prime example for locating half-rooms is using the common space outside several bedrooms for group play or study for households with children

Flexible Floor Plans

Houses designed for the mass market need to provide *flexibility*. In fact, it's always most interesting to visit occupied houses that we have designed to see how the owners chose to use and furnish the rooms—as invariably they've done so in many different ways. In order to be successful, floor plans must adapt to and accommodate diverse types of households.

Plans need to have dual function spaces such as lofts that can either be left open as a multipurpose space or be closed off as a bedroom. On the first floor, living room/home offices might also be able to function as a temporary guest bedroom. Large open spaces can be furnished according to a variety of uses, more so than walled-in and defined spaces.

This idea of "flex space" can be used to make floor plans adaptable to the life cycle of households. The following are some examples:

1. *A small bedroom next to the master bedroom* may initially be a nursery, and later a private sitting room, separate bedroom, or office.

Fig. 5.26 In order to be successful, floor plans should be adaptable to the life-cycle uses of a single house by the same household, as well as to the uses of a diversity of households

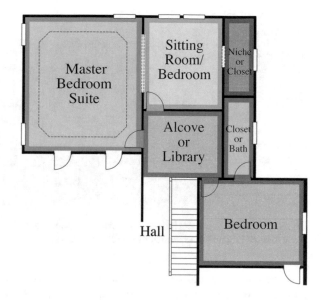

2. *A first floor home office/library* may become a mother-in-law suite for an extended family living situation.
3. *A larger first floor room* with an adjacent full bathroom may be adapted to a master suite when the occupants become less enamored with climbing stairs.

These are just some needs for flexible space that should be considered in designing houses so that a plan can respond to different types of households initially and adapt to the household's needs over a period of time.

Projected Rooms and Sunrooms

One of the time-honored ways of making rooms special is projecting them from the rest of the house—then punctuate the walls with windows. These sun-drenched spaces take on a delightful quality that can easily make them the most desirable room in the house. Projected rooms can be incorporated into even very small homes and can be particularly enjoyable as formal or informal dining rooms, as well as kitchens or sitting rooms. Further, when a room "breaks out" of the building envelope, its ceiling can be vaulted or detailed in a special character.

Projected rooms can be like large extended bay windows. If you think that a bay window might be appropriate for a particular room, then perhaps that whole room should be extended into a bay along the outside wall. Projected rooms do not need to be overly articulated. Often, it is just rectangular or triangular in shape, just large enough to effectively make a departure from the outside wall of the home and develop a special character for the room.

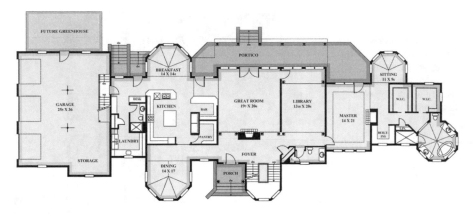

Fig. 5.27 This large custom home uses projected rooms as a design theme. These rooms are much like extended bay windows with glass on all exposed sides

A *sunroom* is a projected room that has extensive glass in the walls and/or ceiling, and sometimes are not heated. Sunrooms are like glass pavilions attached to the outside of a house—they often become the most lived-in room in the home. Historically, sunrooms functioned as screened porches in summer, and unheated rooms with glass panels in the winter. Sunrooms were popular on many different size houses prior to air conditioning. More recently, they have been popular as greenhouse-type solariums for plants, etc.

Fig. 5.28 The popularity of sunrooms can be attributed to their projected presence on the main house, and to their three exterior walls made primarily of glass, as seen in this historical example

Fig. 5.29 In many new floor plans, the kitchen opens directly into the great room and informal dining area. The centerpiece of this kitchen is a very functional work island, with seating, that acts much like a podium

The Contemporary Country Kitchen

One of the biggest distinctions between historical and contemporary house plans is the increased design emphasis on the *kitchen*. Kitchens in smaller, older homes were treated as utilitarian spaces. In larger historic homes, the kitchen was totally ignored because food was prepared by servants and thus not of social emphasis. In new production houses, however, the kitchen is universally acknowledged to be the most important room in the home. Providing the optimal location, configuration, and features in the kitchen is critical to the success of almost any home design.

Perhaps the biggest historical influence on current kitchen design was the *American country kitchen* of farm homes throughout the nation. The country kitchen (where informal dining occurred at a table within the generous cooking space) was the precursor of today's *"eat-in" kitchen*. People's interest in a casual cooking/eating/socializing ritual evolved the country kitchen plan into the kitchen/breakfast/family layout, where in one large space households can spend most of their time. The popularity of this layout can be seen by watching almost any television sit-com, where one of the key sets is invariably the kitchen, laid out for dramatic and comedic exchanges that reflect day-to-day activities in many real-life homes.

In historic country kitchens, counters lined the walls of the kitchen and bench-like seating was in the middle of the room. In the new country kitchen, some counter space is located along the walls for appliances such as the refrigerator and oven, but the sink and/or cooking range is often in a central island counter—which also functions almost like a podium on a stage from which people can communicate with others in the breakfast or family rooms.

The country kitchen is very much the social and nerve center of an operating household. It often includes a small desk or alcove space that will accommodate a laptop computer and places to administer household finances and schedules. The kitchen may weave into the home entertainment area and/or be overlooked from the second level for better communication. It is also handy to be nest to the garage and a door to the outside of the house. A transitional space between the garage and kitchen, commonly known as a "drop zone" to locate coats, backpacks, etc. is also popular.

I will be coming back to the kitchen in Chap. 7 to discuss more details and features found in the new country kitchen. The important thing to recognize in this chapter is how the kitchen relates to the overall plan of the home, adjacent rooms, and the outdoors.

Daylight Basements

One design concept that has sustained validity from historic times to today is the idea of capitalizing on lower level rooms that, due to the slope of the land, can have windows and access to light. These spaces are generally called daylight basements. With the introduction of windows, daylight basements take on the enhanced character of being more usable rooms.

Lower level rooms bring to mind places such as the "rec" room, game room, workshop/hobby spaces, even a lower level bedroom, etc. all of which generally take on an informal character due to their subterranean location. With generous access to natural light, they become far more enjoyable than those that are totally underground.

Some suggestions to make daylight basements more livable and blend in with the rest of the house are as follows:

1. *Detail stairways so that they are easy and inviting to descend.* The typical open-tread, minimal basement stair behind a closed door is not inviting. By calling for basement stairs to match the upper stairs, people will feel more comfortable descending to the lower level. Closed treads, more gentle rise and run dimensions and better handrails can help.

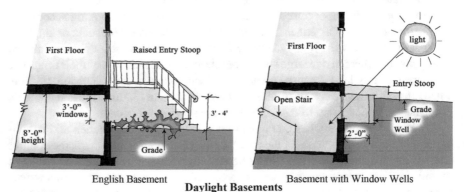

Fig. 5.30 A historically popular design concept is the lower level "rec room," also known as an improved or finished basement. With daylight and improved access from the main level of the house, this space can become even more enjoyable and in some cases also accommodate a bedroom with bath

1. *Open the basement staircase to the upper rooms.* Instead of having a door to the basement at the top of the stairs, consider locating the door at the base of the stairs so that an open stair effect can be achieved on the living level. This may include a partial open railing protecting the stair, which further integrates the stairwell into the surrounding room.

2. *Keep basement ceiling heights similar to those in the rest of the home.* Lower ceiling heights do not make for enjoyable living spaces. Since basement areas are often used for mechanical ductwork, there is an even greater need for increased floor-to-floor height to allow for enclosure of ducts and other utility elements.

3. *Consider raising the first floor height two to three feet above grade to allow larger windows into the basement.* This creates the "English basement" that was popular in urban row houses, such that raised stairs to the first level allowed the lower level to be a habitable space. The same principle applies to production homes. Raising the first floor above grade allows for better window sizes for light and ventilation to the basement. If raising the first floor of the house is not practical, "window wells" can be used to bring light into lower level rooms.

4. *Include a door with direct access to the outside.* An outside door will make lower level basements more functional. This can take the form of a conventional grade-level door in a "walk-out" basement, or it may work with an exterior stairwell in a half basement. For full basements, prefabricated hatchways with steep but functional stairs are popular.

5. *Locate basement utilities so that they can be compartmentalized.* In laying out a basement, remember that mechanical equipment such as furnaces and water heaters should be located in the least desirable place for natural light—and in a place that allows partitions to screen them from the rest of the basement. These walls can be added by homeowners at a later date or can be offered as an option by the homebuilder. When considering future screen wall locations for utility rooms, care should be taken to ensure the resulting spaces are functional.

6. *Consider stubbing in plumbing lines for a future bathroom.* In larger houses, a lower level bathroom or powder room could also be offered as an optional upgrade. This common amenity can also make smaller production houses more livable. The actual fixtures are generally added at a later date so that little additional cost is incurred initially during construction. Plumbing in lower level spaces such as basements may not be possible without pumps to connect to the sewer line.

Work Rooms, Garages, and Storage

Functional rooms of the house, such as the laundry, mechanical rooms, workshops, and storage areas, historically have been in basements or garages. As households move toward addressing practical concerns, they are now the focus of increased design interest. There is a real tendency for designers to ignore functional working

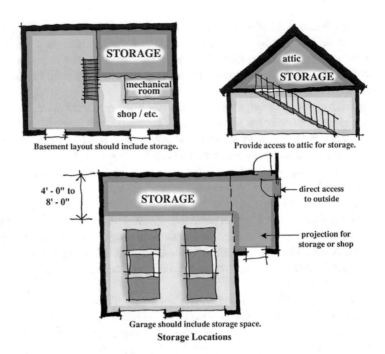

Fig. 5.31 The need for workrooms and storage space is all too often overlooked in the residential design process. Garages with storage components are chronically undersized

spaces because of the strictly utilitarian nature. They are often located and sized as an afterthought, squeezed in where space will allow. Yet to the consumers of housing, the priority is just the opposite: storage and practical concerns are very high on most new home shoppers' lists, with the general attitude about storage being "*the more the better*" and "*you can never have too much!*"

It is easy to make a case for better-designed work areas in terms of livability. Certainly, there is no reason why work areas shouldn't be more centrally located as alcoves off major rooms—as opposed to remote, unlit dungeons in basements or garages. A laundry room that includes space for light crafts, model making, or weaving deserves a location with lots of light and ventilation. On the other hand, certain work functions should be remote due to noise and/or odors such as woodwork refinishing, painting, and the like.

Integral to all homes is the issue of *adequate storage*. A well-designed and livable home will acknowledge that people have things, and they need to keep their things stored. Housing designs must address outdoor storage as well as indoor long-term and short-term storage. Designers need to look for storage potential in every possible dead space in the house. Attics need to be designed for long-term storage with reasonable access via pull down stairs. Some larger homes even merit a full stair to the attic, as seen frequently in older homes. Basement layouts should accommodate short- and long-term storage space with clearly defined areas.

Short-term storage spaces, otherwise known as closets, must also be adequately sized and conveniently located throughout the home. For even a modest home, minimum short-term storage areas include bedroom closets, a bathroom linen closet, a kitchen pantry, and an entry coat closet. For larger homes, we may add front and rear coat closets, hall closets, kitchen closets, hall cabinets, and other short-term storage space. Remember to look for small nooks and crannies created by elements such as stairs and dormers, in order to squeeze every inch of storage space possible into the interior of the home.

In the *garage*, we see other storage issues. The garage typically accommodates a lot besides cars—yard care tools, recreational items, and sometimes mechanical, laundry, and shop areas. In order to fit all these functions into a garage envelope (along with cars), adequate space must be provided at the perimeter of the garage. These may even be projections from the garage wall, with direct access to the outside. Windows or doors with glass panels are also good. The small, cramped, utilitarian garage is another typical space that gets little design consideration, yet is very carefully reviewed by consumers.

Fig. 5.32 There should be a discreet service zone somewhere in or around the house for the storage of trash and recyclable containers. This need is even more important with local governments now mandating large containers that homeowners wheel to the curb to be used

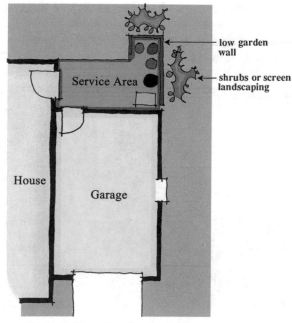

Trash & Recycling Service Area

Trash and Recyclable Collection

One design need that is becoming more and more critical is a place to store and organize *trash and recyclable materials*. A visit to almost any new home community after "trash day" will show how problematic this has become. Many municipalities have mandated that residents wheel a large trash container up to their curb for collection—never mind that years ago collectors would go to the back of the home to collect the trash. Now, in a most uncivil way, our trash stands out in the front of the house so collectors do not have to walk very far. And, human nature as it is, most of these cans never make it back to the back of the house.

New home designs must anticipate trash day with a convenient place, perhaps at the side of the home, developed as a mini-service yard for holding trash cans and recyclable materials. These locations must be hidden from the front of the house in a well-screened place with garden walls or landscape materials from the outside.

Balancing Interior Plans and Exterior Elevations

I have combined interior and exterior design issues in this chapter to make the point that they need to be considered simultaneously during the design process. Interior plan objectives must be balanced with their impact on the house exterior, lot, and street. Conversely, exterior concerns must be balanced with the floor plan needs of the market, such as privacy and affordability. I would like to see this balance return to something akin to our prewar housing, where streetscapes and façades reflected a sense of balanced priorities.

Fig. 5.33 Interior floor plan objectives must be balanced with their impacts on the exterior elevation, lot, and street. In prewar housing, streetscapes tended to reflect a better sense of balance between these priorities, as can be seen at this house in Palos Verdes, CA

Note

[1]Davies, Thomas D. and Kim A. Beasley, AIA. 1992. Fair Housing Design Guide for Accessibility. Washington, DC: National Association of Homebuilders.

Chapter 6
Interior Details

This chapter will continue the discussion of floor plan arrangements at a finer level of detail and look at the smaller elements that make a house more livable and personable.

The standards of interior detailing found in production homes have declined drastically in the postwar years due to general cost-cutting practices. Most interior details were historically site-built by craftsman/carpenters who are no longer active in production housing construction. Their craft is, however, still very much alive in larger custom homes, which continue to have superb millwork installed, though at great cost.

I have recommended to smaller and medium size production builders that many popular interior details should still be made available to their customers on an a la carte basis, priced according to what each optional add-on will actually cost to build, plus an administrative markup. While this concept is generally considered to be too much paperwork and coordination for the larger builders, I still believe that it can be quite viable for homebuilders of more modest scales.

Fireplaces

The fireplace is a perfect example of how "technology mitigation" has been incorporated into new home designs. Fireplaces have been retained in homes over 150 years after their utilitarian purpose was eliminated by centralized heating sources, as a testament to its sentimental and emotional value. Fireplaces are often very much a part of even the smallest affordable homes, and people still tend to think of them as the symbolic heart of the dwelling.

Fig. 6.1 (**a**) Traditionally, fireplaces have been heavy, site-built masonry structures, but with the introduction of (**b**) prefabricated assemblies, designers have the freedom to locate fireplaces in new roles

© Springer International Publishing AG 2017
J. Wentling, *Designing a Place Called Home*,
DOI 10.1007/978-3-319-47917-0_6

Fireplaces were traditionally heavy, site-built masonry structures. Over the last 40 years, however, fireplaces in production houses are now prefabricated metal boxes, allowing the fireplace be located in more creative ways. Most often gas-fired, fireplaces can be used as room dividers or decorative features on almost any wall. Prefabrication has reduced the cost of fireplace construction and given designers more freedom to locate them in places that were once difficult.

Prefabricated metal fireplaces come in various formats—see-through, three-sided, four-sided, and gas "flueless" models (requiring no venting)—and can be used as room dividers or focal points. The most popular location for the fireplace, however, remains where people spend the most time, typically the living or family room. Further, the time-honored location for the fireplace is the formal center-of-the-room spot, with a mantel above and a hearth below.

In larger homes, it is increasingly common to find two fireplaces: one more "functional" for the family room, and one more "symbolic" in the more formal rooms. These roles are frequently reinforced by their detailing—the functional fireplace is generally larger and has a more extensive hearth, mantel, and surround, while the symbolic tends to be more modestly detailed and smaller.

Many larger homes have additional fireplaces located in the master bedroom, study or an outside porch/patio. Fireplaces are also working their way into places where they function as room dividers or occupy corners where they can be viewed from several rooms. As such, they are losing some of their traditional "heavy" appearance from the inside and outside of the home.

Another excellent but seldom-used location for the fireplace is the kitchen/break-fast area. This heavily used spot can allow enjoyment of a fire while eating, during food preparation, and during and after dinner. The kitchen would be a good location for a two- or three-sided fireplace that could also be viewed from the family room.

In custom homes and some production housing markets, masonry fireplaces are still included as a standard feature. In these cases, it is wise to consider both the interior and

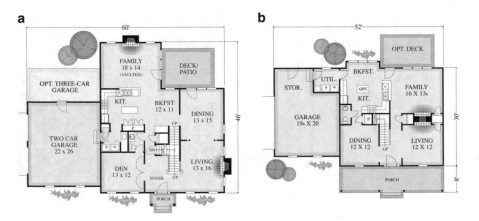

Fig. 6.2 (**a**) Larger houses may have two or three fireplaces, with the traditional fireplace usually located in the family room, and the symbolic fireplace in a formal room or den. Other popular fireplace locations are in the master bedroom suite, kitchen, and even outdoor patios. (**b**) Prefabricated fireplaces—with two or three sides—have allowed the fireplace to act as a room divider and can be viewed from several rooms

exterior impact of the fireplace location on the design—chimneys can help give an elevation both character and value. Historically, many small home designs located the fireplace in the front of the home so the masonry chimney could become a major feature on the street elevation. Today, this is seldom done, because most homes do not have a chimney, only a box-like projection with a vent. When a masonry fireplace is specified, it may be best to locate it on the side of the home where it can be seen from the street, whereas wood frame projections can be better located in the back of the house, hidden from the street, because—let's face it—these "doghouses" look cheesy.

Indoors, a fireplace generally dominates the wall where it is located. To frame the fireplace, locate windows on either side. A common detail consists of bracketing the fireplace with built-in shelving on both sides. The fireplace's interior hearth and exterior chimney have traditionally been places for builders to splurge with precious construction dollars to impart value. Inside, prefabricated metal assemblies are covered with marble, stone, or wood veneer, along with a raised or flush hearth in material of equal quality. Mantels and wood surrounds can include detailed millwork finished to highlight the fireplace setting.

Today, one is more likely to find the media/home entertainment center next to the fireplace; often the fireplace location is subordinated to the flat-screen TV or media center location. I feel it's better not to combine the media center (to be discussed later) and fireplace into one design element, although it is equally wrong to ignore the fact that these two features are usually in the same room, particularly in smaller homes. Further, let's recognize the awesome role of the flat-screen TV as the contemporary electronic hearth. Most household gatherings take place in front of a media screen, not the fireplace. I recommend locating the flat-screen and the fireplace on adjoining perpendicular walls so they can both be viewed from the seating area, but are set apart from one another.

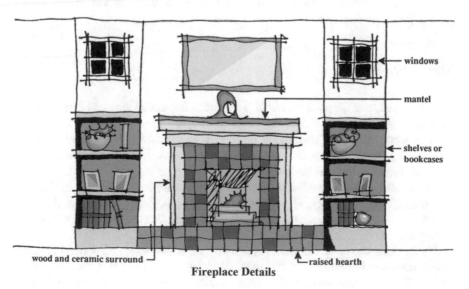

Fireplace Details

Fig. 6.3 The fireplace has traditionally been a feature with extensive detailing surrounding or adjacent to it. Built-in shelving, small windows, raised hearths, and mantels are common details for enhancing the hearth

Fig. 6.4 The media wall or media center is a focal point of new home interiors. Some larger home designs include a special media room as a home theater, but most designs still add a media wall to one of the rooms

Let's also not overlook the fact that some households still use fireplaces for their original purpose to provide heat. Some homeowners in the Northeast routinely use a vented fireplace or wood-burning stove as a primary means of heating their home. In that case, a central location within the home where heat can easily rise to heat auxiliary rooms is optimal.

Media Walls

As mentioned, the fireplace's role as a "hearth" for gatherings has given way to electronic devices. Since the 1950s, households have seen TV and media viewing become significant. Designers have responded by accommodating visual screens and sound systems into media rooms in larger homes. In average houses, media walls are found in lofts, living and family rooms.

The media wall is full of shelves and cabinets to accommodate equipment and recording libraries. Ideally, media walls are made of cabinetry that can close and hide the screen and equipment when not in use—such that the room can be used for other purposes such as entertaining.

Media walls should be located such that they do not consume view walls or dominate a room that has other uses. For example, a media wall might work best along a zero-lot-line wall that needs to be blank, or along a sideyard wall with little view potential. Media walls must work with other design elements (windows, fireplaces, circulation, etc.) such that they do not compromise non-media viewing activities.

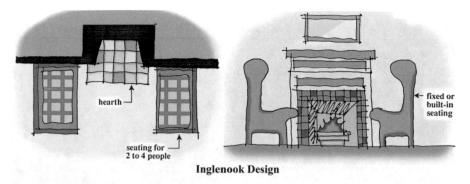

Inglenook Design

Fig. 6.5 The inglenook was a historically popular detail in houses. It acknowledged the role of the fireplace as an occasional mood-setter

Inglenooks

Getting back to fireplaces, let's recall the use of "inglenooks". Fireplaces can be very enjoyable in smaller cozier rooms, historically called inglenooks, that are actually subsets of a larger room. Enjoying the warmth of a fire often requires sitting very close to it. In Frank Lloyd Wright's original home in Oak Park, Illinois, has a small inglenook is found directly off the entry, with built-in bench seating on either side. This approach makes perfect sense for today's role of the fireplace as an element of ambience.

An inglenook does not require a fireplace to be effective. They can serve many purposes where it is desirable to have a smaller alcove off a larger room. One example might be a small space for a desk and telephone that is off a kitchen, living room, or bedroom. The inglenook provides a semiprivate environment that allows one to feel intimate while still being part of a larger space within the home.

Eating Nooks

Eating nooks are another example of a historical detail that has fallen out of favor with production homebuilders because of its site-built nature and extensive millwork costs. Eating nooks were standard in modest prewar homes, located off the kitchen for informal day-to-day dining. The allocated spaces, however, were so small ($5' \times 5'$) that the table and bench-like seating had to be built-ins. While new production homes usually include a "breakfast" area, this is a much larger space ($10' \times 10'$), sized to accommodate a freestanding table and chairs.

Today's breakfast areas lack the charm that the smaller eating nooks offered. Nooks were characteristically cramped but cozy spaces, drawing from the intimacy that small spaces evoke. There's a sense of togetherness that results from dining in close quarters. Homebuilders can still offer eating nooks with built-in seating in production houses on an optional basis. We have recommended this idea for many

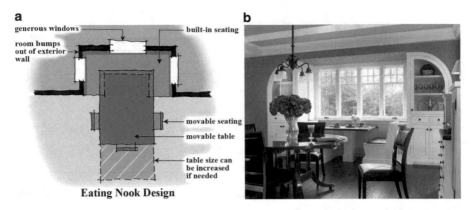

Fig. 6.6 (**a**) Eating nooks can be built with just a small section of built-in seating that, when combined with a conventional table, can accommodate groups of various sizes. (**b**) The area can be customized by the size and style of its furnishings to relate to the larger common space. (Photo (**b**) courtesy Visbeen Architects)

affordable housing prototypes, typically in a space behind the garage and directly off the kitchen. This location coincides with another off-the-kitchen amenity that can be included in the option—a "mud" room with a washer–dryer, laundry sink, and closet spaces that also have direct access to the outside.

Eating nooks can be built inexpensively—a simple counter top table and plywood benches, when appropriately decorated with stencils and cushions, could provide a focal point for a larger kitchen in moderately or affordably priced houses. Good natural light along the adjacent walls is highly desirable; sometimes, the nook space can also be popped out of the building envelope so that skylights can be located above.

Eating nooks can be designed to be flexible and expandable. By constructing an enlarged window seat along a wall near the kitchen, a moveable table can be used for a table top, augmented with additional chairs as required for the household size. This hybrid work/breakfast area reduces the cost of built-ins provided by the home-builder and allows the owner to customize its final size.

Angled Staircases

Another design element that can help personalize a house is the staircase. I addressed some stair design issues earlier—discussing the benefits of having both a front and rear stair. Many affordable homes cannot justify two stairs, but instead may benefit by adding a twist or turn in an otherwise basic "straight-run" stair. Climbing stairs should be an enjoyable experience—and providing a mid-point landing gives the stair climber an opportunity to pause and take in a new view. A landing also adds safety to a staircase.

An angled staircase is probably most cost-effective when it turns 90 degrees at the landing. The added cost of this change in direction includes: (1) additional square footage for the landing and (2) the usual costs associated with corners—extra cuts in the walls, finishes, etc. A 180-degree change in the direction of the

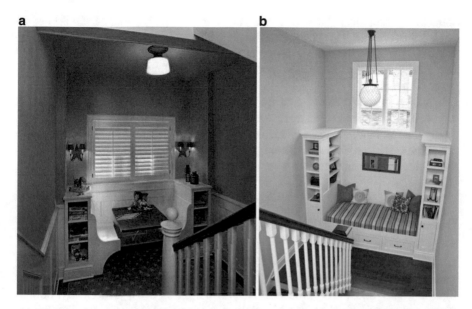

Fig. 6.7 Staircases have traditionally been used to add some drama or charm to a house. (**a**) These stairs with a mid-point landing provides space for an intimate pause, such as with this seating arrangement that could be used for doing homework or a board (or video) game. (**b**) Mid-point landings can also create spaces for a cozy library-type space with comfy seating. (Photos courtesy Visbeen Architects)

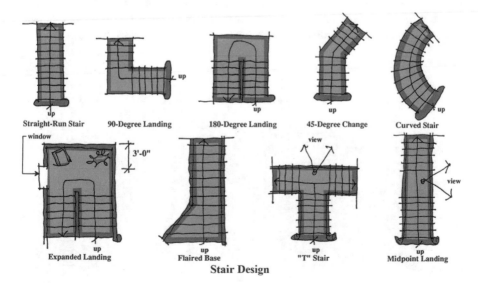

Stair Design

Fig. 6.8 Stair configurations can be angled or curved, and landings can be added to make stair climbing more of an enjoyable and relaxing experience. "Straight run" stairs are increasingly uncommon in design

stair, or "scissors stair," calls for a larger landing—double that of the 90 degree angled stair change. This added cost may be justified by the increased drama of the landing—the scissors stair provides a mid-point area that can actually be expanded into a small usable space, perhaps enough for a piece of furniture or a built-in window seat or shelves.

Changes in stair direction may be less than 90 degrees and still be dramatically effective. Here, you need little more than an expanded tread at the point of change. Without a full landing area, this method of changing direction may, however, present some safety problems, particularly for older users. Care must be taken not to make the angle so minimal that a step may be missed in descending the stairs.

Dormer Rooms and Windows

Dormer rooms are rooms where the ceiling height is diminished in part or all of the room by the slope of the ceiling. Dormer windows break through the roof to provide light and air—and in the process form full height alcoves. Historically, dormer rooms were located in the attic or third floor of larger homes, while smaller designs incorporated dormer rooms on parts of the second floor to save construction costs and/or give the interior and exterior of the house added character.

Dormer rooms are enjoyable spaces due to their altered scale. Lowered ceiling heights on the sides of the room provide a cozy sense compared to standard ceiling heights. Portions of the room with lowered ceiling heights create interior design opportunities for the placement of furniture and wall treatments that can further enhance the character of the room. Dormer rooms are ideal for more flexible spaces such as guest bedrooms, studies, playrooms, and other secondary rooms.

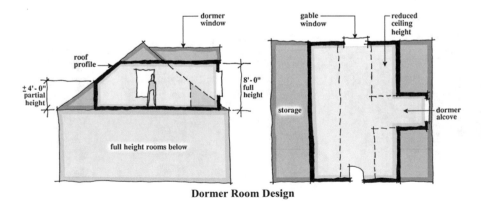

Dormer Room Design

Fig. 6.9 Dormer rooms were historically located in the attic or third floor of houses. Their lower ceiling heights provided an altered scale, generally making the space feel cozier

Dormer windows (which can be located off full height rooms) are also opportunities to incorporate charm into a house design. The scale establishes an alcove that can be perfect for a built-in window seat or small piece of furniture just suited for the space. A double window dormer can easily accommodate a small seating alcove for two chairs, and such an arrangement can work well off of a bedroom.

Dormer rooms and windows are not found in most new home designs, mainly because of the widespread use of prefabricated roof trusses instead of traditional roof rafters. The highly economical roof trusses introduced in the 1950s reduced expensive on-site labor, but also effectively eliminated articulated roofs and dormer rooms. The intermediate vertical cross members of trusses create a space that is unusable except for attic-type storage.

The popularity of roof trusses has put a major damper on the programming of dormer rooms and windows, but there are still some good ways to still include them at minimal additional cost. Usually, the best approach is to use roof rafters on just a small portion of the house, with the majority of the roof covered with roof trusses. Trusses can also be ordered in configurations for dormer rooms. Often called attic trusses, they are designed to have sloped ceilings in areas of reduced head height much in the same fashion as stick-framed dormer rooms.

Fig. 6.10 Dormer windows create alcove-like spaces that are perfect for window seats or smaller furniture

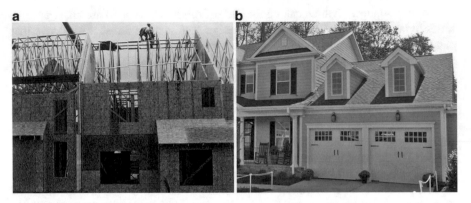

Fig. 6.11 (**a**) Web-like structural roof trusses effectively eliminate the use of dormers in the space within the roofline. (**b**) Rooms with dormer windows over the garage are a popular concept, as here roof rafters or specialty trusses are required only on one smaller section of the house and add to the appeal of the elevation

Bay Windows and Window Seats

Like the projections discussed earlier, windows can bump out of the building envelope to form a small but inviting spaces that easily become the focal point of any room. As these windows reach out to capture more light, they affect both the interior layout and exterior elevation of the home. From the inside, bay windows can add additional square footage to make the overall room feel more spacious and/or create a smaller alcove area.

Bay windows have traditionally been most popular in dining rooms and are often seen in even very affordable homes. Other designs have used the bay window repeatedly as a theme for the overall house. Many large apartment buildings used bay windows extensively to draw in additional light and make small rooms seem larger and special such as the famous "Chicago window."

In production houses, the bay window is an expensive feature due to on-site labor costs. Therefore, bay windows are likely to be used sparingly, if at all, in affordable houses. For smaller homes, however, we still suggest using a "wide" bay window in one room that may span the width of the entire room. This is very effective when the room is tight on space or in need of a feature to enhance its location. The wide bay window can be used in any number of places such as the living, master suite, or breakfast room to expand space and add character.

Smaller bay windows are also found in secondary rooms such as a study or sitting rooms where the scale of the room is intimate. Here, the bay window can either simply add more area or be fitted out with built-in seating. In smaller rooms, a window seat can be more appropriate to give the room charm. Window seats are generally perceived by homebuilders as another one of those site-built millwork specialties where the cost is difficult to justify. In reality, they needn't be elaborately constructed. Window seats can be built from wall framing and drywall with a wood finish board on top. More elaborate versions can be made out of higher grade interior wood and may include trim and storage area below.

Fig. 6.12 (**a**) Bay windows are a traditionally popular design feature generally found in formal living or dining rooms. (**b**) The idea behind window bays can be extended to projecting walls of entire rooms, such as in a second floor master suite or bedroom

Window seats can be incorporated into dormer windows as an inexpensive yet attractive amenity for secondary bedrooms. You can also cut down on the cost of a projecting window by building a "box bay," which is a bay window without the angled walls; instead, the floor below the windows cantilevers out to save on cost. Box bays don't look as articulated from the inside or outside as a bay window, but can be a way of including a projected window feature in an affordable housing program.

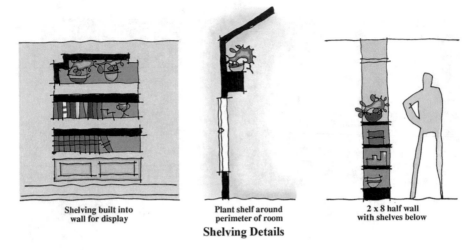

| Shelving built into | Plant shelf around | 2 x 8 half wall |
| wall for display | perimeter of room | with shelves below |

Shelving Details

Fig. 6.13 Built-in interior features such as shelving and moldings were common historical details that have fallen out of popularity due to cutting costs, yet oftentimes these features need not be expensive to construct

Built-Ins, Shelving, and Moldings

Window seats and eating nooks are examples of built-ins. Other items that can be built into an original design may include shelving, cabinets, and trim molding. Many prewar houses included built-in features as common sense and input value. Today's production homes typically do not include built-ins except at the higher end of the market.

One of the most popular historical built-ins has been simple shelving—used for displaying books or knickknacks that virtually every household has. Bookshelves are thought to be popular in a study, office, or library-type rooms, but in smaller homes built-in bookcases might be found in the entry hall or living room as well. Many prewar house designs would include a small alcove-like space with built-in shelving and cabinets to create a museum-like display area.

Display areas encouraged by built-in shelving help to give houses a very personal touch inviting conversation and interaction. What could be more interesting than a display of books that reveal the owner's interests? Or pictures, sculptures, knickknacks, etc. that give a visitor more insight about the host?

Built-in shelves are usually not found in new home designs because they too are viewed as an expensive millwork item. But they don't need to be made out of prime grade wood or in a cabinetry configuration. Shelving can be recessed right into a wood stud and sheetrock wall quite easily and inexpensively by casing an opening similar to that of a door or window and placing the shelves in the wall cavity. Here, the problem is thickness—most walls are made out of 2-inch by 4-inch studs that are really only

3-1/2 inches thick—not a very deep shelf. To solve this problem once a shelf wall is designated, that wall can be made of 2-by-6 or 2-by-8 studs in order to accommodate the shelves. This works well with "half walls," which we will discuss shortly.

Many production homes, particularly in the West, do include built-in ledges around the perimeter of some rooms, which are known as plant shelves or soffits. With the advent of vaulted rooms and high ceilings, builders began to include ledges and soffits to help break up long expanses of walls. These shelves were well suited for plants and other objects that could be enjoyed from a distance. Precedent for this practice is also found in older homes where molding was included around the perimeter of rooms, and was deep enough for the display of plates or other smaller objects. Designers and builders should look for display opportunities to be built-in to production homes inexpensively through cabinets, shelving, and molding.

Columns, Archways, and Half Walls

One of the current design trends noted in Chap. 5 is the shift toward more open plans. Evolving structural components such as floor and roof trusses, engineered joists and composite wood beams have allowed even small houses to economically eliminate walls between rooms. Therefore, it is common to find living rooms and dining rooms programmed as one large space.

One of the reasons for the elimination of walls between rooms is that it makes the resulting combined space look larger and more spacious. The downside of combined spaces is that they can be too casual and barn like. People like some definition

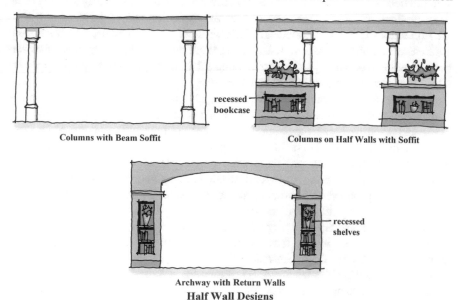

Columns with Beam Soffit   Columns on Half Walls with Soffit

Archway with Return Walls
Half Wall Designs

Fig. 6.14 The shift toward open plans has made partial walls, columns, soffits, arches, and other details more popular for defining rooms without making them feel closed off

between rooms to retain an air of formality to the historically formal spaces. For this reason, designers and builders have reintroduced partial height walls, columns, and archways to "define" rooms enclosing them.

Half walls are commonly used to separate a foyer from a living room—or a living room from a dining room. Half walls may combine 42-inch high walls with full height columns that give the half-wall vertical support. Other times, partial walls may include wing walls with columns located off the wing walls—another historically popular detail. In other cases, elaborate milled columns alone may set off or accent a space.

Half walls and partial walls are excellent places for the built-in shelving discussed earlier. Because they are not extensively long walls, they may be increased in thickness to accommodate shelving. We have often used thicker half walls between rooms and designed "cut-outs" that both open up views and act as display areas. This can be inexpensively done with drywall and some wood trim.

Many historical designs included beautiful turned wood columns as defining points for certain rooms. They have fallen out of use primarily because of cost. Columns can still be included on production designs through the use of "composite" designs of wood studs and drywall. By wrapping two 2-by-4s with drywall and adding wood molding at the top and bottom, a column design can be inexpensively achieved.

Kitchen Details

The kitchen is one area of the contemporary house where historical details are of little application. As discussed in Chap. 5, the expanded role of the kitchen within the home goes beyond the rather utilitarian role of older kitchens, which had few appliances and little counter area. However, there are a few historical ideas that can be reintroduced into new production housing kitchen design. One area is the *pantry*, which was also sometimes known as the *larder*.

Recently, more expensive house designs began to include walk-in pantries. In older homes, pantries were actually small rooms that sometimes even included windows. These had extensive built-in shelving for the storage of foodstuffs. Older pantries were intentionally left unheated in order to keep food fresh from spoilage. The inclusion of the large pantries explains why there were fewer cabinets in older houses—before processed food was the norm, fresh foods needed to be stored in the pantry. The idea of a large walk-in pantry should eliminate the need for some overhead cabinetry, which in turn helps to open up the kitchen to adjacent rooms. With households tending to buy supplies in larger quantities, larger pantries are gaining favor again.

In climates where homes have basement foundations, larger pantry areas are included below the first floor. This defeats the purpose of convenience, but reduces the cost of the space that allocated to storage. Basements are also typically better locations for food storage, temperature-wise. But in climates where housing doesn't include basements, pantries are often in the utility room area between the garage and kitchen, a space that can be half finished and may be less well heated. Still larger homes can include a walk-in pantry right off the kitchen where it is most convenient.

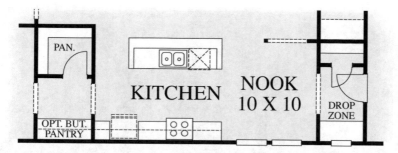

Fig. 6.15 Kitchens are reincorporating walk-in pantries—akin to historically unheated rooms called larders—whose installation can help eliminate some of the need for costly and obtrusive overhead cabinets in the rest of the kitchen. Another popular feature is the "drop zone" between the garage and house interior

Another popular feature in contemporary kitchens is the central island. Older country kitchens had movable work tables, some of which could be placed in a central location with access all around. This is the ancestor of the popular island counter in today's kitchen. One of the reasons island counters are popular today is because they let people socialize while they work. A centrally accessible workspace allows people to work together on food preparation, or engage with each other while working. Island counters are expanding their role as the central focus of the kitchen and can house the main sink or a secondary sink, with different surfaces and levels. They are also now more commonly located to view rooms adjacent to the kitchen, such as the family room and the breakfast area.

Another detail to be observed in older kitchens is an accessible rear door. This direct link to the outside portion of the lot historically led to a rear stoop or porch. Today, the access is more likely to be to the rear deck or patio area. While outdoor dining is the main reason for direct kitchen access to the outside, it is interesting that most new home designs have access doors to the rear from either the breakfast or family area instead of the kitchen workspace itself. The most direct path for bringing out food would indicate that the door be closer to the central part of the kitchen.

Layering of Details

The application of interior details is a process that the homebuilder may start, but that the occupant must finish. Many details likely will be added over time in various remodeling or "home improvement" projects undertaken when appropriate given the time and money required. This process may be thought of as "layering" of details over time, and for this reason many older houses are viewed as having more charm. Homebuilders can start the process of layering by including a certain level of detail that will inspire the future owners to repeat or enhance them in other areas of the house.

Fig 6.16 This advertisement for a turn-of-the-century Sears, Roebuck and Co. house showcases the extensive millwork and built-in features that were standard in house at the time. Today, homebuilders may start the process of detailing interiors, but residents now must often share the role of personalizing their homes over the course of homeownership. (Image courtesy Houses By Mail, copyright 1986 Wiley Books for the National Trust for Historic Preservation).

Chapter 7
Exterior Details

In this chapter we will return to the outside of the house to look at exterior design details in greater depth than I offered in Chapter 5.

To recap: the appearance of our houses suffered tremendous deterioration in the postwar period. The design focus on interior layouts took precedent over the impact of these new floor plans on exteriors. Cost-cutting objectives, smaller lots, and larger garages also took their toll on exterior façades. In many new home communities, houses have lost a sense of character, order, and balance, particularly in the lower price ranges.

Many times the source of the problem is that the exterior appearance of the house is an afterthought in the design process, instead of being integral to the design. In reality, the floor plan is resolved first, and the exterior comes later—a typical directive at the completion of a floor plan is, "OK, let's slap some curb appeal on that." Adding building elements and trim then becomes an exercise after the fact is also an exercise in futility.

Another problem, discussed in Chap. 1, is that the outside of the house is generally not a priority for spending construction dollars. I have often heard this directive from builders: "do something nice on the outside but don't spend any money." The general response to this directive? Mission Impossible.

To improve the exteriors of production houses, homebuilders must assign a higher priority to these areas. Two things must happen: (1) builders must spend more time to consider and sweat the details on the outside of the house and (2) devote a greater percentage of the construction budget to the outside of the house.

In order to develop an integral design approach to exterior design, let's review some thoughts discussed in Chap. 5 that were integral to both floor plans and building image.

Fig. 7.1 The exterior design of production houses suffered tremendous deterioration in the post-war years

© Springer International Publishing AG 2017
J. Wentling, *Designing a Place Called Home*,
DOI 10.1007/978-3-319-47917-0_7

Fig. 7.2 In these designs, the focus of the front elevation is clearly on the entry. A central location in the façade under a covered porch accents the entry door and provides welcoming images for visitors

Establishing a Focus

Typically, the primary elevation of a home will have a focal point. This is frequently in or near the center of the façade, and most often involves the entry. The focal point could be a large projection on one side that includes an important room, or it could be a turret on a corner that is curved or angled, or it could be a two-story portion of a primarily single-level plan.

Some houses may have more than one focal point—possibly two or even three. Other designs may have a major focal point and a secondary one. It is important to establish at least one focal point however, and then reinforce it within the overall exterior design. Let's illustrate this issue with the most common focal point of a front elevation: the entry door.

I had discussed accentuated entries in Chap. 4, and listed design ideas to reinforce the entry area in detail. The following are some additional suggestions to make the entry a focal point within the overall façade:

1. *Dress up the door.* Using sidelights, trim, and headers around the door can help provide a focal point for the overall house at little added cost.
2. *Push the entry wall out.* An entry wall in front of the rest of the façade wall has more visual prominence; conversely, recessing the façade here can downplay the importance of the entry.
3. *Locate a gable element over the door.* This may be flush with the front wall or project from the façade. It can also incorporate a covering over the entry door.
4. *Call for a specialty window over the door.* Specialty windows can also be used on either side of the door in the entry gable.

Another common focal point of a façade other than the entry could be a portion of the plan that projects forward and creates a strong massing element. This projected area could be enhanced with a variety of elements including the following.

Fig 7.3 In this house, the focus of the front elevation is a large specialty window above the entry door

1. *Locate a specialty window on a section of the façade.* This may be larger or different in character from other windows—such as a bay window, "Palladian" arrangement, or "picture" window. The window design may include half-round shapes, a transom, or a different muntin pattern.
2. *Implement a material change that contrasts from the balance of the house.* Add quality veneers such as brick or stone to further enhance the focal point's prominence on the façade.
3. *Add distinctive detailing around focal point windows.* Flower boxes or special trim are effective choices to make a specialty window into a focal point.
4. *Add a decorative element in the gable or on the roof.* A special vent or attic window, for instance, can provide additional articulation over the focal point.

After focal point(s) for the façade have been determined, our discussion on detailing the exterior will now go straight to the top—the roof. Let's start there, and then work our way down to the base of the home.

Stylized Roof Profiles

Historically, roof profiles vary from region to region based on climatic conditions and the adopted architectural response. The roof profile of a house is frequently the most distinguishing feature of a particular design signature. For example, the hip roof makes us think of Mission or French eclectic houses, while steep side-gabled homes make us think of the Tudor style. Low-pitched roofs with deep overhangs are

Fig. 7.4 Stylized roofs are more common on (**a**) houses with roof rafters, or *stick framing*. (**b**) Roof trusses tend to make houses more box-like in appearance

typical of bungalow and craftsman homes, while the gambrel roof is characteristic of Dutch Colonial houses.

Stylized roof profiles continue to be seen in custom homes and there are some examples in production homes. Builders can still incorporate different effects such as steep pitches, varied overhangs, and complex hipped roof designs, even with roof trusses as the primary roof structure.

Trusses vs. Articulated Roofs

I mentioned in Chap. 6 that one of the new building technologies that has had the most impact on house exteriors is the prefabricated roof truss. The design impact is significant. Roof trusses consist of a series of connecting wood members that create a web of lightweight wood. When placed over the walls of a home, the interior area of the truss becomes dead space, eliminating the possibility of attic or other usable rooms permitted by roof rafter construction.

The widespread use of roof trusses has tended to reinforce the box-like look for affordable homes. When floor plans are reduced to the least common denominator, a rectangle, with roof trusses above, you can see the worst-case scenario for housing design—a mundane cigar box with a shallow roof pitch. Since any jogs in the plan and roofline add to the building cost, the incentive to keep affordable houses in these simple box-like configurations is great.

The impact of roof trusses on the design of larger homes is less apparent because of the market-driven need to have a more articulated plan and façade, but trusses continue to move the elevation toward simplistic rooflines. Roof details such as dormer windows, partial walls, hip roof profiles, and other breaks in the roof add cost and therefore are discouraged.

In spite of the limitations of roof trusses, residential designers can develop strategies that will allow some degree of articulation and character while still capitalizing on the cost benefits of roof trusses. Ideas for accomplishing this include the following:

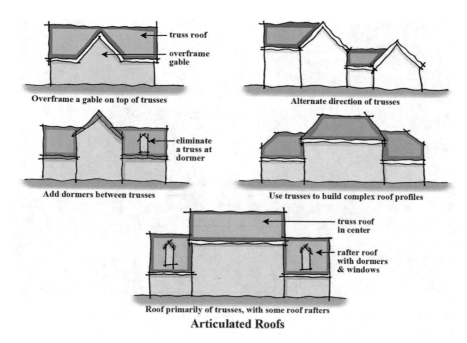

Articulated Roofs

Fig. 7.5 Roofs built with roof trusses can have articulated designs through a variety of techniques

1. *Frame most of the roof with trusses, with rafters specified on certain sections.* We commonly design roof rafters over the garage so that dormer rooms and windows can be incorporated. To further achieve cost savings, these areas may be stick-framed as predetermined options that buyers select and purchase in advance. Therefore the builder does not risk putting extra construction dollars into a base design that may cost more than a competitor's basic model.

2. *The most common way to mitigate the negative look of a long truss roof is to "overframe" a gable area on top of it.* This type of gable is built directly over the top of the roof and creates a sort of "double roof" with even more dead space. Overframing, however, can relieve a long section of roof and create the impression of a more articulated plan and elevation below.

3. *Dormer windows can still be built on roof trusses so as to act as skylights to rooms below.* This approach calls for either using narrow dormer windows that fit between the trusses (which are usually spaced two feet apart) or for skipping a truss to create an opening wide enough for a window to be placed on top.

4. *Consider providing variety in the roof profile by switching the direction of some or all roof trusses.* Most homes are designed with roof trusses spanning from front to back walls, or perpendicular to the length of the house. When the truss direction is changed (to side to side), it creates a very different look for the same house plan. In larger homes with several roof truss profiles, the common practice is to run all of the trusses in the same direction. But by alternating the direction of trusses at one of the roof sections, variety can be achieved on the façade at a minimal cost.

Fig. 7.6 This very affordable entry-level model house in Durham, NC incorporates a dormer on top of a vaulted truss roof. The narrow dormer here acts as a skylight into the living room below

5. *Consider a combination of trusses and roof rafters to create an articulated effect.* A hip roof, for example, can be engineered with most of the roof designed in trusses and only a minor section in roof rafters. Truss manufacturers are able to provide articulated roof profiles when needed for higher priced homes since they are always seeking market expansion.

Low vs. High Roof Pitches

There is a tension in residential design between the savings effected with a shallow roof pitch versus the negative visual impact of shallow pitches on the elevation. Historically many house designs utilized the attic or area under the roof as usable space, and steeper pitches were justified. With the introduction of roof trusses there was no practical reason for roof pitches to be steeper than necessary for water drainage.

Roof pitches are expressed in terms of rise against run. A 4-over-12 pitch (4/12) rises 4 inches in a 12-inch length. An 8/12 pitch is twice as steep and a 12/12 pitch is three times as steep. Generally, the minimum roof pitch required for the use of roof shingles or tile is 4/12, which, when viewed in elevation, appears very shallow. This is the pitch used on most manufactured homes where cost savings and transport issues govern the roof design, as well as on many site-built affordable homes. As the price of the home increases, typical roof pitches generally increase to 8/12 or even as high as 12/12. In roof pitches, height generally connotes value, prestige, and quality.

Fig. 7.7 Though a common cost-saving measure, a low roof pitch can often make a house look *too* cheap

In eastern styles, we advocate roof pitches on even affordable designs to be in the 8/12 range. If roofing spans are reasonable, as discussed in Chap. 5 under *Narrow Modules and Spans*, then the cost impact of steeper pitches will not be much greater than for shallow pitches. For deeper spans, steep pitches may be prohibitive unless the area under the roof is utilized in the design concept.

Another method of incorporating steep roof pitches is to designate that any over-framed gables or other modules with gables facing the street or other prominent location be steeper in pitch than the balance of the roof. For example, the main section of the roof that spans from front to back may only be 6/12, but a small gable section facing the street may be 12/12. This sets up a hierarchy between utilitarian roof pitches and more decorative ones.

Recall that roof pitches vary historically and regionally, and that a shallow-pitch roof is not always undesirable. Western Mission style homes, bungalow homes, and craftsman designs successfully incorporate shallow pitches as a design motif for climatic reasons. Conversely, styles such as English Tudor and Cottage have very steep (over 12/12) pitches as a design characteristic. These roofs are integrated and justified with overall design of the house.

Roof Overhangs, Trim, and Fascias

The relationship between a roof and the balance of the home is another area of detailing that has been largely omitted in new homes. Historically, this was a transitional point in the structure that bears some analogy to the treatment the ancient Greeks devised for their temples, where columns would have elaborate capitals supporting decorated lintels. Many house styles emulated these classical orders with brackets, fascias, and subfascias to accomplish a strong transition between the walls and roof.

Fig. 7.8 Roof overhangs, trim, and fascia boards at the intersection of the roof and walls are often overlooked detailing opportunities, and builders of modest houses may only be able to justify these more elaborate transition details on a front gable or wall

Roof overhangs are regionally and stylistically varied for climatic and environmental reasons. Western ranch style homes and bungalows are known for deep sweeping overhangs while New England homes usually have minimal overhangs. As a general rule, houses in colder climates have steep pitched roofs with minimal overhangs, while homes in the south typically have shallow pitches and deeper overhangs (12 inches or more). This is consistent with the role of overhangs—to help block sunlight and heat gain during warm months, while allowing light and radiant heat to penetrate during winter days.

Some designs such as Dutch Colonial homes have exaggerated eave overhangs, which help identify the style. The large eaves were originally implemented for practical purposes of keeping rain and snow away from the home. Today these same themes can also be used to create a sense of protection in locations where needed, such as in front of entry doors or projecting windows.

Roofs have two types of overhangs. The roof *eave* is the overhang at the wall running parallel to the roof ridge. The roof *rake* is the edge created when a wall intersects the roof perpendicular to the roof ridge. In most cases, the eave overhang is wider for practical reasons (rainwater runoff); the rake overhang is more of a decorative detail. Shallow-pitch roofs may look more substantial if featuring pronounced eave and/or rake overhangs, as typically seen on craftsman homes, while steep pitches may eliminate the need for an improvement with the rake condition.

Other design elements that help accentuate the house-to-roof relationship are the fascia and trim detailing. A fascia board is the piece of wood that ties together the individual rafters or truss members, which is very prominent on any elevation. The thickness of the fascia board is another design issue that should be carefully considered in addressing the overall design theme for the home. A narrow fascia board (6 inches) can look cheap, while a wider fascia (8 to 12 inches) will look more prominent. Some fascia designs can be enhanced by adding a "subfascia," which means there are two layers of boards that make up the overall fascia assembly. A subfascia enhances the look of the roof trim by giving it more thickness. Builders of more modest homes may only be able to justify a more elaborate fascia design on the entry façade or at a front gable one that does not need to run the entire perimeter of the home.

The other roof trim element that has been almost universally dropped from affordable and mid-priced housing is the frieze board. The frieze board is a trim piece that is attached to the wall right below the roof, forming a transition between the wall material and soffit, or roof overhang. In Georgian and Federal styles, the frieze was a prominent element, articulated with additional dentil molding patterns. Today friezes are primarily used on historical replications of period styles, though I suggest the concept can be appropriate on all designs.

Dormers, Partial Walls, and Roof Windows

I have already noted how dormer windows on roofs can add interest to an elevation, particularly where large expanses of the roof plane occur. Taking a look at older homes, the use of vents, windows, and partial height walls enlivens the roof and gable elements around the home. These ideas are particularly important to consider in light of the large, unbroken roofs that are seen in higher-density, small lot houses discussed earlier.

Upscale builders can enliven the roof by adding traditional dormer windows that, while simply adding light to an unusable attic, nevertheless look good from the outside. There is significant precedent for doing so, although most roof and gable windows, such as the "eyebrow" window, are more admittedly decorative. Other examples of

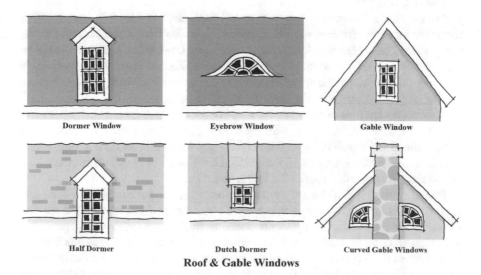

Roof & Gable Windows

Fig. 7.9 Special windows, vents, and other design elements should be considered so as to break up long expanses of monotonous roof and wall gable areas

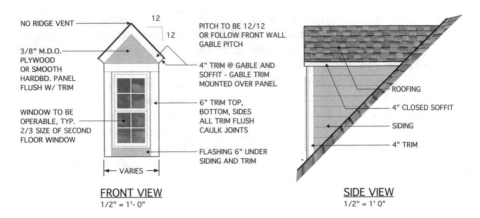

Fig. 7.10 Adding extensive details to dormer window elements helps maintain correct proportions and styling throughout the façade

decorative roof fenestration include smaller square or circular windows, half-rounds, and quarter-rounds.

Decorative windows frequently appear in front gables of homes, where they are most prominent and least expensive to install. Less commonly are they found on top of roofs, where additional expense is required to frame a roof to cover the window. In spite of their cost, we sometimes recommend smaller dormer windows be placed on the roof, particularly when there is a lower section of the roof where the dormer will be very prominent. Ideally, there will be vaulted space below the dormer that allows the window to be functional and let light into the room below in a fashion similar to a skylight.

We also have designed homes with small decorative windows in gables where the gable faces the front of the lot. These need not be functional windows in the conventional sense, but they can certainly enliven a large gable area, which may be a focal point of the elevation. Many historical residential styles had small gable windows that ventilated the attic. While today's attics are ventilated by other means, smaller gable windows can still serve a decorative purpose.

Another design idea to articulate the exterior roofline is the partial height exterior wall that creates "half-dormers" windows. This historically common detail would call for the exterior wall to be partial height—say 4 to 6 feet above the floor. When windows are located on the wall, they would have to be built as half-dormers, and from façade they would "break" the fascia board line.

Partial walls and half-dormers were originally used to save money on construction materials. Today the opposite is true. As you can probably guess, the rigidity of manufactured design components such as roof trusses and wall assemblies makes partial height walls expensive departures from the norm. But much in the same vein as dormer windows, partial height walls can be used judiciously to provide visual interest on the façade as well as interior character. Partial height walls can be effective in visually "bringing down" the fascia of a large roof at some portion of the elevation.

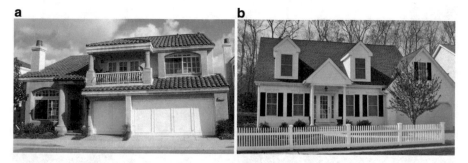

Fig. 7.11 The two predominant "skin" materials found on most new houses are (**a**) stucco in the West and South and (**b**) siding of vinyl or fiber cement in the East and Midwest

Building Skin: Stucco vs. Siding

Moving down from the roof to the walls of the façade, we now come to detailing exterior materials. Typically production homes have two types of skin: (1) a primary, utilitarian material that covers the majority of the home's outside and (2) one or more accent materials used in limited quantities due to cost constraints.

Almost all houses built in America use one of two primary exterior materials. Houses in the East and Midwest use siding, while houses in the South and West are primarily stucco. Why? Because they're the least expensive game in town! Historically stucco came to dominate certain Western regions because wood was not as readily available, but today stucco is also used in the North and Midwest when the market calls for it. Here it is generally an upgrade material, because it is more expensive than siding.

Of course, today's siding is not wood—it looks like wood but is actually vinyl, fiber cement, or a wood substitute product developed for the low-maintenance exteriors demanded by the market. These products, while eliminating the tri-annual paint job, have other visual negatives, such as buckling and heaving, and joints that give the façade a "cheap" look. Siding manufacturers have been working to remedy those problems and are making significant progress in the design of their product.

The stucco material of the West is a superior exterior material from a design perspective. Being an East coast architect with some design experience in the West, I am often envious of the freedom stucco allows. Stucco is a forgiving material. It can be sprayed over a built-up wood frame forms to give exterior walls dramatic sculptural details in a very cost-effective way. Further, stucco can be given an infinite number of color variations, from home to home on a street or within an individual façade.

Designers and builders should become comfortable with maximizing the appeal of these skins, particularly on affordable homes where accent veneers are not likely to be in the budget. In my opinion, stucco is the easier material to master—the key is to master the interplay between light and shadow created with wall projections. Stucco homes can often be successfully detailed with "projections" —windows,

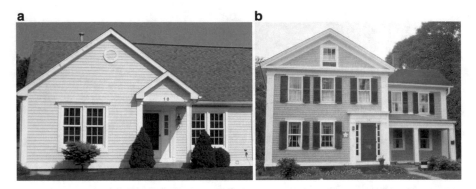

Fig. 7.12 (a) A typical Eastern Berkus house design includes bold transitions between walls, roofs, and windows. (b) This practice can been seen in historical designs, such as at this Greek Revival house in East Longmeadow, MA

window surrounds, niches, planter shelves, and the like that can be carefully composed on an exterior wall to create a pleasing effect on the façade.

I have always felt that one of the most successful production home designers, the late Barry Berkus and his firm, Berkus Group Architects, were helped along and inspired by the stucco clad Neo-Spanish architecture of the firm's Santa Barbara, California location. In translating the historical architectural lessons found in this Mission town, Berkus was able to use built-up wood forms covered with stucco to recreate the arches, arcades, pillars, and other details on even modest new home construction.

A hallmark of Berkus designs was a simple but important design thought—make all forms and surrounds bold. Stucco pop-outs and wood trim molding are much thicker than what is generally seen in residential designs. The Berkus Group design signature of bold, direct forms gave their homes strong visual interest and appeal. (See Photo 4–20 for an example.)

On the East coast, architect Bill Devereaux, formerly of the Berkus Group, had to translate their philosophy to building skins made of siding. Much of the same transitional boldness was implemented with heavy, wide trim pieces terminating the siding at corners and around windows. Where typical corner trim is 4 inches wide, Berkus designs will have trim 8 inches or wider. Often trim areas around window forms are so wide that trim material was not available so metal-clad dimensional lumber was substituted to achieve the bold effect.

With stucco or siding, using bold forms around the windows, doors, and other projections appropriately defines the siding and clearly expresses transition points at windows, doors, and corners, particularly so with the use of trim. This approach can be further reinforced with color changes between siding and trim.

One can see similar inspirational detailing of siding on older homes in New England. In this region, it is quite common to see very wide trim boards at corner, window surrounds, frieze boards, and base boards. The reason for this is that wood siding was used and was readily available in a wide variety of widths. Today's vinyl or fiber cement trim material is available in very limited number of dimensions.

With respect to siding, we can take heart. There is significant progress going on with product development to overcome the negatives. Manufacturers are anxious to tailor their products to more sophisticated design requirements. Siding companies are improving colors, trim, sizes, and profiles that will give designers much more opportunity to achieve high quality designs.

Integral vs. Applied Veneers

Another heartbreaker in terms of the prevailing design standard for exteriors is the generally poor locations selected for veneer applications. The too-common norm is to put a quality veneer such as brick on the front façade only, with the rest of the home abruptly changing to siding or stucco. This reinforces the image of cheapness that plagues new home design. It is a shame, as construction dollars for these expensive veneers could be used elsewhere more effectively.

Veneers look best when they appear integral to the structure and not applied as an afterthought. For example, we favor locating masonry along the base of a home at the foundation, where it can appear as a strong, solid foundation for the structure as opposed to just one vertical section of wall. Another theme is to develop a masonry wainscot where the veneer is carried up several feet above the foundation before transitioning to another material; masonry as a heavier "base" material is appropriate.

Perhaps only a portion of the house such as a one-story component should be veneer, making a clean break with the other form. Transition is the key phrase in veneers. The challenge is to make a graceful transition from one material to the other so that the veneer does not look pasted on. This favors veneer locations at inner walls and under roof projections, places where the veneer can terminate cleanly and logically.

Where do these veneers make sense? Walls below overhangs are a good place to use a masonry material; it visually makes sense to see stone or brick here. Pillars and walls that are used to support something above again make sense, as that is what stronger materials do—provide structural support. The common use of a masonry veneer along the foundation also makes sense creating a "pedestal," which sets the house above the ground on a strong-looking base.

Fig. 7.13 (**a**) Veneers that appear integral to a design are more appropriate than those that look merely applied. (**b**) Masonry used on the facade of a house looks bad when it appears to be pasted only onto the full height of the front facade only

Where do veneers not make sense? How about places where they cantilever over a wall of lighter material? Or a two-story section of wall instead of an equal amount of one-story wall? Or how about a section of wall around a chimney, but not the chimney? These are common blunders that I have seen that seem to contradict the basic nature of veneer materials.

Veneers should be located where they have the most appropriate visual impact, typically on the front façade. They should be where they can be appreciated, such as at eye level, near the entry, or at the base of the home. And be careful not to use too many different veneer materials. As with colors, two or three are fine, but beyond that additional veneer materials can create a mish-mash effect on the elevation.

Window Shapes and Window Dressing

Windows typically consume about 15 % to 20 % of the total construction budget, more than any other building component. Therefore, specifying window types and detailing are critical to the success of a well-designed house.

Locating windows again recalls the discussion of balance—what will look great from the inside of a home may look inappropriate from the outside, and vice versa. The composition of fenestration elements needs to be considered throughout the design process, and not just when developing the floor plan.

One of the most common mistakes in window design is the selection of too many different window types. Historical residential styles often had a stylistic window theme. For example, we think of a Tudor home having somewhat narrow casement windows, or a Dutch Colonial as having wide, double-hung windows. Homes in the Mission style are recessed deep into the wall and may have iron grillwork on the outside. Mixing different window styles and sizes within a single design is visually confusing.

Fig. 14 Window design can establish or reinforce a design theme for the entire house. This French Country style house calls for tall and narrow casement windows throughout the entire plan

Fig. 7.15 Correct detailing of windows is critical to the success of a façade. (**a**) Shutters that are out-of-proportion to the window sizes are problematic. (**b**) Well-received window dressing techniques may include added headers and sills, trim, shutters, and flower boxes

How many different sizes are appropriate? I would go with no more than three window sizes on any façade. These can fall into (1) major windows into prime rooms such as living rooms; (2) secondary windows into other less important, but still well-used, rooms such as bedrooms; and (3) small windows for bathrooms, utility rooms, and attics. This combination of windows might translate more literally into a bay or "Palladian" window assembly in a key first floor room, conventional double-hung windows in most all other rooms, and small windows in bathrooms and other rooms that need just a little light or ventilation.

Often a "specialty" window such as a half-round, circular, or large fixed glass window is specified on the front façade. Avoid having too many specialty window shapes, lest they compete with one another. One special window per façade is enough. Place them so they are indeed part of a focal point on the elevation, instead of being crowded in somewhere as an afterthought.

Whether specialty or conventional, windows must still fit into the overall window rhythm of the façade and home. As discussed in Chap. 5, window placement is a critical part of the design process for establishing a sense of order within a façade, and need to be located in a way that creates visual balance.

Beyond the selection and placement of windows, we also find a major opportunity to enhance the look and feel of a house through window dressing.

Windows have historically been embellished with treatments such as shutters, decorative trim, awnings, and flower boxes on the outside, muntin patterns on the window glass and curtains or blinds inside. All of these elements, some functional, some decorative, help enliven window openings that connect the outside and the inside of a home.

Fig. 7.16 (**a**) The inclusion of muntins in windows helps to develop human scale in the façade. (**b**) One problem with "snap-in" muntins is that some residents remove them, making some houses look unfinished when viewed from the street

I have had extensive debates with my architectural colleagues over the use of some of these elements, particularly in reference to nonfunctional shutters—which manufacturers have reduced to flimsy pieces of plastic that end up nailed to the wall on either side of windows. The use of these phony tack-ons goes against the grain of many designers' sensibilities, and for understandable reasons.

I accept that these elements are indeed nonfunctional. However, in many housing types they offer an important opportunity to add a splash of much-needed color and decoration to an otherwise drab façade. Besides, almost every culture since antiquity has included some form of decoration on their architecture: the Egyptians carved stone columns to look like bundled reeds, Greeks' Corinthian capitals were stone acanthus leaves, etc. I see window shutters as a legitimate form of decoration, along with some of the other nonfunctional elements seen on new homes; the fact that they are simply decorative is of no consequence.

I do hate to see decorative elements used in inappropriate ways. Shutters should be sized as they would be in their historical role: to cover the window. The use of small shutters on either side of a double window looks very inappropriate. Ideally, they should be proportioned to the adjacent window and be one-half the width of the window.

Other decorative elements used on the outside of windows include headers (lintels) and sills, sometimes with a decorative design on the header. Window trim is particularly important on housing designs that do not include shutters or other decoration. Although standard trim is usually only 4 inches wide, I recommend bolder widths such as 6-inch trim in many cases. Windows without substantial trim cheapen the entire façade.

Some other more elaborate decorative elements for windows include wrought iron or flower boxes. Although flower boxes are rarely installed by builders due to being such personalized items, we have successfully specified them on an unlikely housing type—rental apartments—and they were very well received by the occupants. Contrary to the usual arguments that people won't take care of them, we found that people not only maintained them, but also undertook additional plantings around their units.

Before leaving the discussion of windows, I will touch on two issues of window design that impact the exterior appearance of the home: muntins and insect screens.

The popularity of "fake" window muntins is another excellent illustration of the high-tech/high touch phenomenon, where manufacturers responded to the public's interest in technology mitigation. Over the past three centuries, one could approximate the age of a home by the number of panes of glass found in the window sash. As the manufacture of glass became more sophisticated, it was possible to increase the size of the glass and reduce the number of vertical and horizontal muntins that were required in the sash. Eventually, the glass panes became available in sizes large enough that muntins dividing the sash were not needed at all.

Today's windows are now manufactured with "snap-in" or other artificial muntins. Homeowners prefer the appearance and charm of the older windows that had muntins, especially since the muntins divided the glass into more human scale elements. While on the whole this was a positive response for window manufacturers, there were some problems from an application standpoint: when muntins "snap-in" they can break, people take them out to clean the window and don't put them back in, or the builder never puts them in to begin with. These situations can lead to presentation of an uneven appearance from the outside of the home when one window doesn't have muntins and the others do, or one home in a community has them and others don't.

This problem can be avoided by specifying muntins that are integral to the glass. Since most windows are now manufactured with double-paned glass, the muntins can be placed inside the panes in a sandwich-like solution. This eliminates the problem of missing muntins, but color selection is limited.

Another contemporary window problem involves insect screens. When these screens are installed on the outside of windows, window details become muted. Some window manufacturers are addressing this problem by placing the screens in a roller on the inside of the window, where they can be pulled down when needed and be less visible from the outside. Other types of windows, such as casements, already have the screen on the inside of the window frame.

Detailing Façade Projections

On historical houses, projections such as porches, bays, and other window elements are generally focal points because of their extensive detailing. These areas can really enhance a façade when columns and/or arches are added—bold elements articulated with smaller trim.

Columns, arches, and openings on building projections are important opportunities to enhance elevations of even modest houses. We generally recommend that posts be upgraded beyond minimum sizes of 4 inches by 4 inches with trim boards that will give them a 6- or 8-inch nominal dimension. Beams at porches and other projections need to be expressed so that you get the feeling the roof is adequately supported.

Railings present another opportunity to provide something beyond the minimum detailing. A decorative railing design such as an altered pattern to the bars, posts, or top rails can be specified at little additional cost. Proper detailing of even just a small area of the façade can stand out and significantly improve the façade as a whole.

Fig. 7.17 Detailing the façade with appropriately scaled elements is critical. Beams, columns, railings, window trim, and other materials need to be sized to impart a sense of stability and permanence to the home

Color Selection

Here is a design issue where, for no additional cost, a home can look very good, or with the wrong colors…very bad. Unfortunately, color selection is often a major downfall of many new homes and communities. Here's why: many builders give their customers free rein in the selection of colors on exterior materials. While this may make sense in satisfying the needs of each individual buyer, it may not be the best course of action for the overall look of the community. Some people just don't have a feel for color.

In other cases, builders prepare standard color combinations for buyers. These designate which colors will be available for stucco, roof shingles, trim, etc. Here we can have the same problem because generally no one is designated to be responsible for color selection. The process often falls into many different laps—interior designers, manufacturers, real estate agents, homebuilders' spouses, buyers, etc. Although we are discussing a highly subjective topic where individual tastes vary tremendously, I feel most color selections on new homes need guidance.

To be specific about the problems and potential solutions, let's divide a home into its color components: (1) roof, (2) exterior skin, (3) trim elements, and (4) decorative accent elements. Here are some rules for each:

1. *Roof.* Look for colors of strength, typically dark greys or browns that one may associate with the lasting quality found in traditional roofing materials, such as slate

Fig. 7.18 Choosing appealing colors for the materials and other elements is tremendously important, and costs literally nothing to enhance the façade. Light color siding with dark accents and a red door works well

or tile. Most roofs are asphalt shingles in the East, and composition tile in the West, which can be manufactured in any color. You will find more of them in deep or dark colors than light ones, although the light colors are more practical because they reflect heat better. I have heard many buyers complain that light colors such as tan and light blue look "cheap" and undesirable. On the other hand, in the southern states such as Florida, light roof colors in pastels of blue or green or pure white are regionally appropriate.

2. *Skins.* On the exterior building skin, the preference is for lighter colors, because the skin color forms a backdrop for the overall house. It also covers the greatest area and could easily become overwhelming to the eye. Perhaps the most popular skin colors are whites, tans, light greys, and very light blues. Use of deeper colors with light trim colors is also valid. Regional variations such as the colors of the desert—muted earth tones that respond to the surrounding desert landscape—are also regionally appropriate.

3. *Trim.* The trim color around windows, rooflines, corners, etc. is usually a light color, preferably lighter than the skin color, or the same as the skin color. White is an ever-popular trim color. In some instances, such as Neo-Spanish designs, dark trim sets off the lighter stucco and warm roof color. Regional variations are found for all general rules.

4. *Accents.* Doors and shutters are where we suggest strong primary colors—reds, blues, greens. Careful selection ensures these colors will be strong when compared to the balance of the home. With light skin and trim colors, the accent materials need to have a particular punch.

The most common problems in color selection include the following:

1. *Too many strong colors on skin, trim, and accent elements often don't work together.*
2. *Accent elements are too weak to distinguish themselves.*
3. *Skin color is too strong and competes or overcomes accent colors.*
4. *Too many different colors (which may stem from too many materials).*
5. *Weak roof color, which can be worse than a strong skin color.*
6. *Mixing warm and cool colors.*

What about selecting colors for a community? We have often been asked how much color variation is appropriate to enliven a streetscape without promoting a sense of disorder. My answer: variation is desirable within a philosophy of color selection. The above general principles are one philosophy of color and color combinations. There are other ways of looking at the issue, and so long as a sense of order prevails, the more color variation, the better.

Detailing for Street Presence

The exteriors of our houses need to regain the individuality and presence found on older homes. Exteriors are perhaps the most important area where a reconsideration of historical building practices is worthwhile. Detailing need not be expensive, as one treatment for affordable houses may include nothing more than consideration given to massing, window placement, and colors. While added construction expense may be minimal, quality exteriors require more time on behalf of the homebuilder to address issues like color coordination, window specifications, and details.

Without more time devoted to these considerations, production houses will continue to reflect a lack of caring for the greater community and individual house. The outside is where designers and builders need to sweat the details!

Fig. 7.19 The exterior is arguably the most important area of exterior design, in terms of considering historical design precedents, addressing color coordination, selecting window specifications, and detailing

Chapter 8
Multifamily Housing

In this chapter we will address the planning and design of multifamily or attached housing in the context of its suburban setting. This will include multifamily of a density that would be compatible with single detached homes, such as the townhouses and garden units found in traditional towns and contemporary planned communities. The discussion will be limited to walk-ups, meaning buildings without elevators that are two to three stories in height and can be built at densities up to around 20 units per acre.

Multifamily housing has some of the best—and some of the worst—examples of all housing design. Historically, one thinks positively of urban and suburban row homes and duplexes, while remembering the negatives of nineteenth century industrial tenements and later failed urban public housing projects. In the current suburban development context, the most common example of multifamily housing may be the anonymous apartment complexes with box-like buildings ringed by parking lots that dot the fringes of metropolitan centers.

Perhaps a reason for the polarization of design lies in the two primary roles of multifamily housing. There are two types of multifamily products: *lifestyle* housing and *affordable* housing. In lifestyle housing, units are built in an attached mode because the setting is made valuable by an amenity, such as a beach, golf course, mountain range, or proximity to cultural and employment centers. What is being sold is as much the lifestyle of the environment as it is—or perhaps more so than—the physical shelter. Lifestyle housing frequently serves as a second home and/or income property that may only be used seasonally. Many lifestyle houses are designed for the luxury market and have generous design and construction budgets.

Fig. 8.1 (a) These late nineteenth century urban row houses in Lancaster, PA include front porches and differentiated façades, which is in contrast to (b) many newer suburban "townhouses." Multifamily designs suffered from the same postwar cost-cutting mentality that impacted detached housing

© Springer International Publishing AG 2017
J. Wentling, *Designing a Place Called Home*,
DOI 10.1007/978-3-319-47917-0_8

As for the other role of multifamily—affordable housing—units are attached to increase density while saving on construction dollars in order to deliver homes at costs below baseline prices for single-family homes. In some instances, multifamily fulfills this role as rental housing, condominiums, or fee-simple ownership. Some designs can be very compact for small households, while other plans can also be quite large and even include garages for the family market.

Here we will focus primarily on multifamily housing as a vehicle to deliver *affordable* homes—housing that meets the needs of smaller households such as first-time buyers, small families, retirees, shared housing, and single-person households. For the most part, these groups need primary housing for their full-time residence.

Urban Vs. Suburban Setting

Imagine yourself as someone from outer space who has just landed on Earth, right in the middle of a multifamily housing community. After opening the hatch and looking around, you scratch your head and say, "I don't get it. Why are these people living so close together? I don't see anything so special about this place!"

a

b

Fig. 8.2 (**a**) These early twentieth century mews houses in Redondo Beach, CA were arranged in a courtyard theme. (**b**) Most garden flats are now arranged in two- or three-story buildings surrounded by parking lots

Indeed, the problem with multifamily in suburbia is that when you take historical urban models such as row homes and garden apartments out to the country, they look like fish out of water. Without the urban street grid, these structures are lost in too much space.

One primary reason people have traditionally chosen to live in multifamily housing is for the urban experience—the ability to live in an environment where you can walk to work, to shop, and to recreation, without the need for a car or extensive travel time. Once you take away those urban amenities, what's the sense of living so close together?

That question drives the argument behind the traditionalist planning movement: that suburban higher density housing must include urban or semi-urban amenities such as quality open space, pedestrian environments, convenient commercial and retail services. Without those amenities, higher density housing becomes something that people will live in because it's the only thing they can afford, when they would really prefer to live elsewhere. Most commonly, they would rather have the American dream—a freestanding detached home with a yard.

Of course there are exceptions to that prior statement. Many people are opting for multifamily housing over single detached homes because attached homes are more likely to be offered with a homeowner association that will take care of maintaining the outside of the building and grounds. An intense dislike of house and yard work or little time for it, along with lower prices and increased security, is a particularly driving force behind the retiree and empty nester acceptance of attached housing, in both for-sale and rental categories.

For other households, multifamily is regarded as the first rung on the housing ladder, bought with hope of "trading-up" to a detached home at a later date. Under this scenario, the objective of multifamily planning and design should include recapturing some of the benefits of the urban setting—the quality open space, the convenience, the amenities—with prototypes that also offer maximum privacy, identity, and livability.

Multifamily Prototypes

Early postwar suburban multifamily was dominated by two prototypes—the vertically organized townhouses and the stacked garden apartments, or flats. Townhouses were based on the historical model of the row house, while stacked flats were simply apartment units built low to the ground with balconies and stairs instead of elevators. The other lower density multifamily example would be the duplex, or twin, that is really somewhat of a hybrid between detached and attached housing, which is why they are often referred to as semidetached homes.

As multifamily became more and more common in suburban communities, planners and designers experimented with variations on basic prototypes. As I explored in a 1988 book I coedited, Density by Design, members of the American Institute of Architects Housing Committee attempted to catalog some of these trends wherein duplexes were becoming manor homes, townhouses were becoming attached singles, and breezeway apartments were becoming courtyard buildings. While the point of that book was to exhibit successful higher density housing, here I would like to focus on planning and design issues that address multifamily as a component of vibrant, mixed-density communities, regardless of building type. So let's start the discussion of multifamily with community design. What are the really important community design issues to consider with respect to multifamily housing?

Address Vs. Anonymous

The threshold design issue to address in designing multifamily housing is the address. Think for a moment of how you would describe where you live. Do you have a number on a street, like 56 Elm Street? Or do you live in Building 3, Unit 3 A? Of course, many prefer the former, because it gives a home a uniqueness and an identity.

Identity is as key an element for multifamily homes as it is with detached homes. It's very important to find the front door of a detached house, and the same issue applies to multifamily. With attached homes it may be even more important to give people a "real" address because the closeness of the homes mandates a greater attention to identity.

There are several approaches in the design process that can help multifamily dwellings establish individuality:

1. *Identify the front door*. In our multifamily designs, whether townhouse or stacked flats, we attempt to give each home a front door that is clearly visible from the most public side of the building. For stacked flats, this will typically result in an internal, individual stair for each second-level home, instead of common stairs and hallways. In some instances where very aggressive affordability/higher density objectives are needed, common stairs will be necessary, but when possible, individual stairs are preferable.
2. *Keep entry doors apart*. Try to avoid the side-by-side entry doors common with attached housing. People do not like to have their doors confused with someone else's. This can be tricky to achieve in multifamily designs, because doors do need to be somewhat near one another—it's the nature of the beast. Try to separate them, however, with low garden walls, separate porch stoop design, and other techniques that identify each door as an individual entrance.
3. *Articulate each individual façade*. Clearly identify where one unit ends and the other begins on the exterior façade. Vertical changes in floor heights (where topography permits), horizontal shifts in the façade, roof breaks, and material changes are examples of differentiation that can occur at the party wall to set units apart from one another. For stacked units, vertical and horizontal changes can include cantilevered floors, material or color changes, and alternate window patterns.

Fig. 8.3 Attached houses generally need to be set apart from one another. A clearly individual front door, separated from the adjacent houses, is optimal. Changes in window and door patterns, as well as material and color changes, will also help differentiate each house from its neighbor

4. *Vary the design of the end unit.* Attached units in a row are punctuated with an "end" unit at either end. This is an opportunity to introduce a significantly different floor plan that will capitalize on the additional exterior wall, which in turn should result in a different façade, roofline and massing from the interior units. It is helpful to have the end-unit roofline lower than the interior units to step the building height down at the ends of the building.
5. *Alternate the window and door pattern.* Be careful: too much variation between units will result in a haphazard look, while just a small amount of difference may help introduce some much needed relief into a long façade. Sometimes a dormer window may switch to a roof gable, or an entry may have a different porch element or door design from one façade to the next.
6. *Look at color changes.* Tricky. Historic row houses were usually built with identical window and door patterns with only the color changes on doors and shutters providing the variation. In other cases where the door and window patterns may vary, restrained color changes between units may be preferred.
7. *Add details that enhance human scale.* Many of the features mentioned in Chaps. 4 and 7 that can help articulate an entry or façade are helpful to incorporate on the façade of a multifamily house. This may include garden walls, planters, quality materials and trim molding, lighting, mailboxes and house numbers, to name a few.

Less Is More

There's a saying in the army. "If you need a squadron, send in a platoon. If you need a platoon, send in a battalion. If you need a battalion, send in a division." You get the point—deal from strength. The reverse applies to programming for multifamily housing. If you don't need elevator buildings, use walk-ups. If you don't need walk-ups, use townhouses. If you don't need townhouses, use duplexes. Typically, a lower density prototype is best for promoting community design and livability. It keeps people close to the ground.

Duplexes, for example, have all the advantages of single detached homes except for one party wall. The floor plan has three sides available for access to light, ventilation, and outdoor orientations. It is easier to articulate the façades and entry doors from the adjacent one. Further, it is easier to make the outdoor space more private from the adjacent unit.

Townhouses, the next higher density prototype, typically offer access to the outdoors on two ends of a plan. They don't have people living on top of one another, thus reducing privacy and noise problems. From a community design standpoint, they can be individualized vertically from the outside as belonging to one household and can have a well-articulated entry statement.

With townhouses, there is a secondary issue to consider: the building length. How many homes are attached in a typical building? Here the same rule applies; less is more. For example, if you add one home to a duplex you have a triplex, or a building of three townhomes. A fourplex would have four houses, etc. Obviously the fewer houses lined up in a row, the better. A triplex has two end units and only one interior unit. A fourplex has two end units and two interior units.

The building length issue, however, must be balanced with density objectives. The fewer units you line up in a row, the less you will be able to increase density. Certain planning objectives can make long buildings an asset, however, such as the crescent block in London's *Regents Park* discussed in Chap. 3.

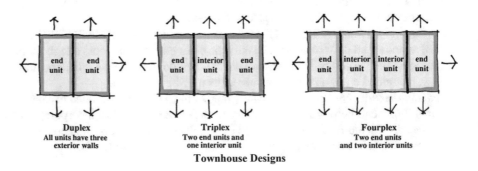

Townhouse Designs

Fig. 8.4 When arranging attached houses into buildings, generally the fewer units per building, the better. End units are typically more desirable than interior units, because they have at least one additional exterior wall that gives access to natural light and air

With garden flats, a number of things can be done to enhance the individuality and privacy of each unit, in spite of the fact that people are living on top of one another. Earlier on, I had mentioned the advantages of individual entrance doors with private, internal stairs for stacked units. Likewise, the entry door for garden flats can be enhanced much in the same way as for townhouses, where porches and other features vary from one unit to the next to set one apart from one another.

There are some other community design issues to consider with garden flats. Besides the length-of-overall-building issue, there is the question of whether the floor plans are "back-to-back" or "through" designs. Back-to-back flats are "backed up" to one another, so that there may be two or even three party walls per plan. These plans are typically used in breezeway buildings with common entrances and stairs. Through flats, on the other hand, have a "front" and a "back," similar to townhouses, because the plan goes "through" the building to have a front and rear orientation. Through flats have community design advantages because they can more clearly orient to a street, and have distinct front/rear orientations. Through flats, however, will typically not be able to achieve the same level of density that back-to-back flats can.

In higher density multifamily prototypes, hybrids of townhouses and flats can work successfully to increase density while maintaining good community design standards. We have frequently designed buildings with townhomes as interior units and through flats as end units. This allows the stacked units at the end of the building some outdoor orientation alternatives to mitigate the over/under deck/patio. It can also provide alternatives to side-by-side entrances.

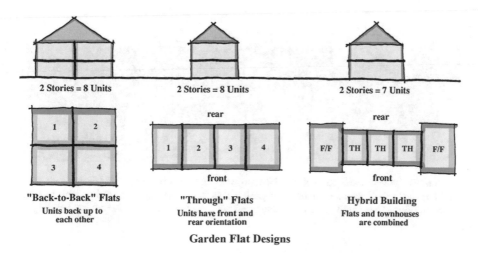

Garden Flat Designs

Fig. 8.5 Garden flats can be organized into a variety of patterns depending on density demands. Generally, "through" plans (plans with a front and rear orientation) are more livable. Garden flats can also be combined with townhouse plans in the same building, to increase density or livability

We have also designed some buildings with stacked through units in the interior and townhouses at the end. This is justified by locating the higher value plan, the townhouse, at the end location for a higher price, which will allow the interior plans to sell at a lower, more affordable price.

From Our House to Manor House

There is another philosophy of multifamily design that applies the opposite approach to the individualized, articulated model. That would be the *manor house* approach, which assumes that in some cases you may want to have a collection of smaller units appear to look like one large grand home, or manor house. Under the manor home philosophy, individual entrances are still possible, but may be part of a larger grand scheme such as a courtyard or grand stair.

The manor home approach works well with multifamily buildings that need common stairs and balconies, because it still gives the overall building a sense of address and presence. Typically, manor homes have both front and back façades and can still allow for individual entrances and outdoor courts.

With the manor home approach it is possible to mix stacked units with two-level townhomes in one building. We have completed several manor home designs that feature one-level plans on the first floor with two-story plans above. With this prototype you have grade-level entrances for housing that is three levels in height, allowing significantly higher densities.

Fig. 8.6 The *manor house* concept may be appropriate at higher densities. This design for condominiums in South Carolina combines ten garden flats into a building intended to look like one larger dwelling. The central gable in the courtyard acts as a focal point for the overall building

Three-Story Townhouses, Stacked Townhouses, and Walk-Ups

While two-story multifamily buildings are preferable to three stories, land costs are frequently calling for higher density designs that work best with three floors. The three-story townhouse is one example that has evolved from this, primarily to reduce the building footprint and thereby consuming less of the site. In the three-story townhouse, the first floor is typically a garage, entry foyer and a first floor room that may be a bedroom or study, etc. The first floor steps lead to a second floor, which becomes the main living area with kitchen, dining, and living spaces. The bedrooms are typically on the top floor. There are variations to this arrangement; however, it is the most common layout.

The main variant of three-story townhouses is the garage location. Ideally the garage is located in the rear of the plan accessed off an alley, allowing the entry door to be off a primary street or walkway resulting in a pleasant front view of the building. In other cases, due to site constraints the garage must be located in the front of the plan next to the entry door, a less than ideal scenario but design concepts may be incorporated to play down the presence of the garage. In both cases, a second floor rear deck is a popular amenity giving occupants some outdoor space off the main living floor.

For "stacked townhouses" of three and sometimes even four stories, basically there are two dwellings one on top of the other between the party walls. Here it is common for second floor unit to have one flight of steps to their unit, sometimes even two flights of steps. There are many variations of how the two stacked units are integrated and how garages may be provided. One concept is to have the second floor split between the lower and upper units, so that both have two floors and a garage. In other cases the ground floor may be a smaller, more affordable or accessible unit, with the second and third floors devoted a larger and more expensive unit. By using design concepts seen in Fig. 8.10, decks and balconies can be located in the front and rear of the building for maximum privacy.

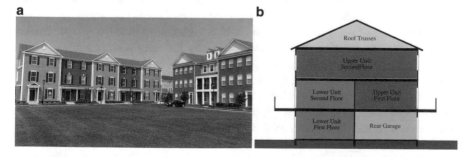

Fig. 8.7 (**a**) These three-story buildings with rear garages face a common green. The building on the left is townhouses, and on the right is *stacked townhouses*. (**b**) With creative design, stacked townhouses can have private garages and private balconies, amenities associated with lower density building types

Fig. 8.8 (**a**) The very common three-story *breezeway* building combines open air circulation with surface parking to deliver affordable housing without elevators. (**b**) There are many variations of walk-up multifamily buildings; this example combines three units in a 40-ft square footprint and has integral garages

At even higher densities, yet avoiding an elevator are three-story walk-ups, most commonly seen as garden apartments. Here typically there is no garage included in the building so surface parking is used for cars with common hallways for circulation. Perhaps the most common is the breezeway building with centrally located open air circulation. For this, four units per floor or twelve unit buildings are idea as all units have two outside walls for views and air circulation.

There are many variations of the walk-up buildings, some with integral garages, some with internal common hallways, others use the slope of the land to have one side with two stories and the other three. A popular configuration for these buildings is the courtyard plan, with entries off a common court than can be attractively landscaped as a common area for residents. The common thread is typically these are affordable housing prototypes without elevators that can increase costs significantly.

The Automobile Factor

With multifamily development, as with single family, the toughest community design issue is accommodating automobiles. For higher density multifamily buildings, the most successful resolution of this has been to locate parking underground or above ground in disguised parking structures. But for the densities we are addressing here—walk-ups—automobiles will be accommodated either on surface parking lots or in garages, which are generally attached.

Let's talk about surface parking lots, where tension resides between the market-driven interest in having cars as close as possible to the door versus the community design need to remove cars from directly in front of the building. Some planning approaches seek to ease this tension:

1. *Side lots.* I suggest locating the parking to the side of, or between, buildings. This allows the front of the buildings to face a street and sidewalk, or open space. The distance from the front door to the parking must still be reasonable (less than 100 ft), but it should be removed from directly in front of the building.
2. *Remote lots.* In other cases, we can pull parking away from being situated immediately in front of the building such that open space or a courtyard surrounds the entrance area. The objective is only to pull the cars away from the area directly in front of the building—not so far as to be an oppressive walking distance.

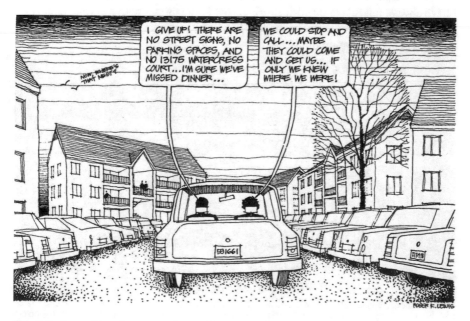

Fig. 8.9 As architect/cartoonist Roger Lewis reminds us, accommodating automobiles in multifamily settings is often difficult to achieve gracefully. (Courtesy Roger K. Lewis)

3. *Park across the street.* In lower density multifamily prototypes such as townhouses, we have frequently located the parking near the front of the homes, but across the street from the entrance, so that cars are not directly in front of the property. With this approach, homes can still step down toward the sidewalk and street. If some of the spaces must be located in front of the homes, they should at least be separated from the front door by landscaping.

4. *Dispersed parking.* When surface parking is implemented more for affordability than for density objectives, if more parking can be dispersed and integrated with landscaping, then the overall community will look better. Large concentrated parking lots are unsightly, so arranging parking in smaller groups as opposed to large lots will mitigate the presence of cars on the site.

5. *Reduced parking ratios.* Many zoning ordinances have excessive parking requirements for multifamily that may be totally unjustified for the market being served. For example, some ordinances require two or more parking spaces per unit without respect to the number of bedrooms within each unit or the occupant profile. Don't be afraid to present evidence from reliable sources on how much parking should be required; ask for a reduction for your project if warranted. Ratios of 1.5 or 1.1 spaces per unit can be more appropriate, depending on unit size, occupant profile, and access to transit. Extra pavement benefits no one.

When a multifamily program is serving a market where garages are desirable, designers need to look at building designs that combine garages with some surface parking to satisfy both resident and guest parking requirements. Attaching a garage to a multifamily home is one of the most difficult exercises in multifamily design. I have always felt that an historical urban row home model with a garage tacked on in front looks ridiculous, but is among the most common multifamily prototypes found in suburban communities.

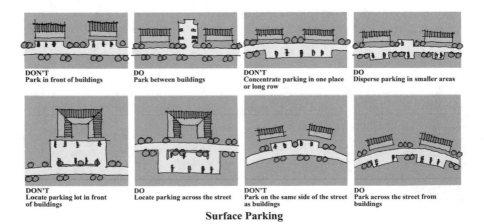

DON'T Park in front of buildings   **DO** Park between buildings   **DON'T** Concentrate parking in one place or long row   **DO** Disperse parking in smaller areas

DON'T Locate parking lot in front of buildings   **DO** Locate parking across the street   **DON'T** Park on the same side of the street as buildings   **DO** Park across the street from buildings

Surface Parking

Fig. 8.10 Parking for multifamily housing should be pulled away from buildings and dispersed. Parking spaces should be located within reasonable walking distances to entrances of home, with acceptable distances varying according to regional climate and market standards

Mitigate the problem of attaching a front-loaded garage into a typical townhouse configuration with some of the following design guidelines:

1. *Limit the garage width to one-half of the overall façade.* Following this rule, townhomes with garages in front should be no less than 22–24 ft in width. This leaves sufficient room for a gracious entrance and some windows allowing light in through the front wall. Garages typically are 12 ft wide, but can go down to 11 or 10 (tight!).
2. *Pull the garage "into" the front façade.* By pulling the garage into the envelope of the plan, its intrusion into the streetscape is reduced. How much should it be pulled in? As much as one-half the total garage length (typically 10 ft) is recommended, however it is possible to implement designs where the front of the garage is flush with the entry façade.
3. *Locate second floor rooms over the garage.* It helps to locate a room with windows over the garage to draw the eye away from the garage door. This also enlivens the streetscape with additional windows that can be "dressed up."
4. *Use quality materials at the garage façade.* A high quality garage door design with window panels and extensive trim is helpful, and the addition of veneer materials on the adjacent walls is also desirable.
5. *On end units, consider side-loaded garages.* On end-unit townhomes, you may be able to implement side-loaded garages. This concept is similar to a single-family home with side-loaded garages—just locate the garage doors on the side wall and bring the driveway away from the street. This will provide some much-needed relief from the front-loaded garages of the interior units.
6. *Integrate some townhomes without garages in the building.* In some programs, you may be able to include a mixture of townhomes with garages and some townhomes without garages, to help improve the street view of the façade by

Fig. 8.11 Incorporating attached garages into multifamily buildings is a challenging design task. Garages should be recessed into the façade, as was done at *Belmont* in Suffolk, VA built by Chesapeake Homes

reducing the impact of garage doors. This will also allow some more affordable models to be offered in the community.

7. *Consider putting the garages in the back of the building*. This idea uses the traditionalist approach of planning alleys in the back of the buildings, removing the car from the entire front façade. This has historically been a popular concept in prewar multifamily, which has not been used in most contemporary planned communities—largely for the same reasons that alleys were unpopular with single-family subdivisions.

8. *Provide detached garages away from the front of the house*. This simple solution puts garages in independent structures dispersed throughout the site, yet near townhome entrances. These smaller structures can be effectively used to enhance the site as a privacy buffer or screen between adjacent uses or buildings.

Some other solutions to accommodating automobiles include the mews arrangement (all entrances face a garden court and garages are along an interior alley) and platform parking, where townhouses are located on top of a concrete deck. These and others are generally for much higher density programs than described here.

Private Outdoor Space

Another major challenge in multifamily design is providing private outdoor space for each unit. Just as you want to allow some separation between entry doors for individual unit identity—the same need applies to outdoor patios and decks—they need to be secluded so that residents will have a sense of privacy when enjoying the outdoors.

The typical solution for townhouse units is to locate the patio or deck in the rear of the plan, in a fashion similar to detached homes. Here screening devises such as dense landscaping or garden walls may be desirable to provide privacy between the units. The narrower the footprint, the more there is a need to include privacy barriers. The following considerations can help to make outdoor space more private for attached housing plans:

1. *Distance*. For wider units, plan to keep a reasonable amount of space between the deck/patio areas of adjacent units.

2. *Privacy walls*. For narrower units, consider screen walls or dense landscaping. These should be as low as possible while still providing privacy—say about 6 ft. Often the height of the wall can be decreased as distance from the outside wall of the unit increases. The screen wall design should have some detailing to mitigate the negative visual impact of large blank walls.

3. *Varied orientations*. End units can have their open space orientation on the sidewall, to totally avoid the issue of being next to the adjacent unit. This may require providing privacy from the adjacent end unit on another building.

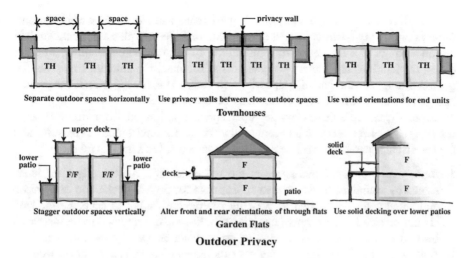

Fig. 8.12 Providing some privacy for outdoor spaces is challenging with attached housing, but there are many design techniques that can aid in this effort

While townhouse units challenge the designer's ability to provide private outdoor space, stacked flat units are even more difficult. Here you have the "stacked deck" arrangement, where outdoor space typically consists of two or three small decks stacked vertically. To address this problem:

1. *Stagger the upper deck and lower patio locations.* This may mean that on one floor the lower patio is located off the living room and on the upper unit it is off the kitchen, or some other combination of rooms.
2. *Alternate front and rear orientations.* For through unit designs, consider locating the lower patio in the back of the plan and the upper deck in the front of the plan. In this scenario, the deck could also provide a covered porch over the entry below. In some cases where units are built on steep slopes, designs can accommodate grade-level patios for both plans.
3. *Provide solid decking materials.* If decks must be stacked over patios, make the deck as solid as possible (no open spaced planks) and consider recessing part of the upper deck to minimize the lost light on the patio below.

Multifamily Open Space and Amenities

One of the elements of multifamily housing that makes it a desirable housing alternative is the concept of shared open space and amenities that are not usually available in detached home subdivisions. For young singles and retired couples, these may include a common pool and clubhouse area, where residents can enjoy recreational amenities without direct maintenance. Other common amenities include tennis courts, volleyball areas, exercise facilities, etc.

The placement of active and passive open space and amenities within a multi-family community is critical. Most often, these facilities are placed near the project entrance to act as a marketing tool for selling or leasing the community. In other cases, the facilities are centrally located. In still others, the amenities may be dispersed throughout the parcel. Depending on the market location, resident profile, and geographic location, any of these may be appropriate.

Typically, there are three ways that open space is addressed in a multifamily land plan: (1) as a buffer between adjacent land uses, (2) for environmental reasons, and (3) for visual separation or active use by the residents. Let's look at all three.

1. *Buffer Open Space*. Open space at the perimeter of a multifamily site, determined by zoning setbacks, is also appropriate for private deck/patio orientation and some active uses such as walking or jogging trails. As such, buffer land should be sized to accommodate these functions and may need to be increased beyond zoning minimums, depending on adjacent land use, view, etc.
2. *Environmental Open Space*. This may include wetlands, steep slopes, wooded land, view corridors, or man-made environmental open space such as storm

Fig. 8.13 The lower patios at *Bentley Ridge* in Lancaster, PA are open to the sky, thanks to the upper units' decks being located off of another room in the plan—giving both upper and lower units more privacy in their outdoor spaces

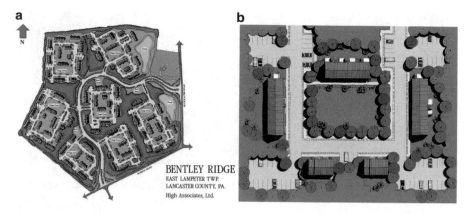

Fig. 8.14 The land planning concept for *Bentley Ridge* was (**a**) to organize the site into neighborhoods of approximately 50 homes each, and then (**b**) to provide each village with a usable common green space

water retention basins. In most cases, this type of open space is also beneficial for private deck/patio orientations, but may also be considered for community open space and oriented in the site for maximum benefit by all residents. This may mean that a retention pond is part of the edge of the site that can be viewed by all, including the neighboring community.

3. *Usable Distributed Open Space*. This type of open space includes the residual land after the buildings, parking, roads, and active amenity/service areas are subtracted. In many instances, it is not distributed thoughtfully and represents a lost opportunity to introduce real character into the community. Depending on the community size, this open space should include some passive greens and perhaps some active fields for ball games, etc. For maximum visual benefit, passive use space should be located near the buildings, and active use more remote from buildings, perhaps near the perimeter of the site.

We see these three types of open space quite clearly at *Bentley Ridge*, a rental community in Lancaster, Pennsylvania. Here usable open space is distributed throughout the community in the form of "greens" that allow each village of roughly 50 homes to have a social open space. Each village is part of the larger 420-unit community that includes perimeter buffer space with pathways and a large retention pond at the community entrance. Here each type of open space has a distinct functional role.

The location of active open space such as pools, clubhouses, and tennis courts, is another issue. Active amenities often include noise and odors that may warrant their separation from living units. On the other hand, amenities such as pools may have a positive view quality, with some units benefiting from a "poolside orientation." Some general rules regarding active amenity location:

1. *Mitigate the impacts of sound from active amenities*. If noise is a potential problem, add buffers between the amenity location and living units. Landscape materials may help address this concern, but there's no substitute for distance.

Fig. 8.15 Buildings at *Bentley Ridge* overlook common greens and sidewalks rather than parking lots

2. *Connect the amenities to the community.* A centrally located amenity area is generally best, and should include a walkway system that connects it to the balance of the community, so that all residents can benefit from the amenities.
3. *Consider the "town center" analogy for the amenity area.* The amenity area should allow for socialization, congregation, and events within the community. Elements such as bulletin boards, clocks, etc. are desirable.

Combining Single-Family Houses with Multifamily Buildings

A major planning and design trend discussed earlier in Chap. 3 is the movement toward integration of different types of housing within the same community. This allows for more resident and architectural diversity similar to the traditionalist model: small town and villages. There is a shortage of successful new communities of this type, since the standard practice for decades has been to keep different housing types completely separated.

New communities being built under planned community ordinances are taking advantage of the flexibility of these regulations to mix single- and multifamily homes within the same site. As the single-family market moves toward accepting and even preferring smaller lots, some of which may even have condominium-type homeowner association and maintenance agreements, the ability to mix housing types seems even more logical. Mixed-housing communities can offer tremendous benefits for buyers to select their desired housing type within a planned community with common amenities or as the traditional stand-alone house and lot alternative. Also mixed housing will allow households to "move up" within the same community without having to uproot themselves from neighbors and schools.

The mixture of housing types within a community needs to sustain good planning principles. What are some of the issues that come into play when singles are built next to townhouses and garden units?

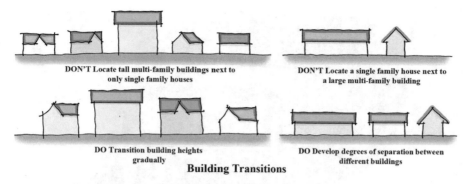

DON'T Locate tall multi-family buildings next to only single family houses

DON'T Locate a single family house next to a large multi-family building

DO Transition building heights gradually

Building Transitions

DO Develop degrees of separation between different buildings

Fig. 8.16 When mixing single- and multifamily housing, consider the visual impact of transitions between buildings of different scales, widths, and heights

1. *Develop degrees of separation.* Should you have a single house next to a six-plex? Perhaps, but probably not. But would they look OK if they were across the street from one another? This siting process should consider transitions from one level of density at a time. For example, singles would be sited next to duplexes, duplexes next to triplexes, etc.
2. *Building heights and lengths should transition gradually.* Using common sense—it's not a great idea to put a small ranch plan next to a three-story garden walk-up. Locating housing types in gradual steps between height and length is more appropriate.
3. *Adopt architectural controls.* Develop a consistent theme within the community for setbacks, driveways, open space, architectural controls, etc.
4. *Develop density rings.* To the extent possible, minimize the need for lower density residents to be impacted by the higher density homes, i.e., having to drive through the entire site to get to the most expensive homes is not a good idea. Nor is it great to have lots of traffic volume past the larger, more valuable homes to get to the smaller ones.

Many of these are a part of *Summerfield at Elverson*, a mixed product planned community in Elverson, Pennsylvania discussed earlier in Chap. 3. As a 200-acre addition to a small village, *Summerfield* was planned to work as an extension of the historic town, which is a collection of single homes, duplexes, and commercial buildings dating back to the Revolutionary War.

New architectural prototypes at *Summerfield* are compatible with historic Elverson, both in design and planning approach. Detached homes have garages in the rear of the lot, porches in front. The initial phases include both triplex and five-plex townhouses, duplexes and single-family detached homes—all within a 50-acre parcel. Homes are related to an open space system that includes pastureland as a theme, complete with a stable for boarding horses.

SUMMERFIELD
AT
ELVERSON
PHASES I - III

Fig. 8.17 *Summerfield* in Elverson, PA is a mixed-density, mixed-housing community with townhouses, twins, and single-family houses sited altogether on the same streets

Integration of Commercial into Multifamily

One of the goals, or dreams, of traditionalist planners has been the idea of reintroducing mixed-use buildings into suburbia—residential over commercial. This historical concept of "living over the store" has tremendous merit, but equally daunting challenges to implement in contemporary planned communities. Most commercial developers are unromantic about residential over commercial and are focused on the need for commercial to have visibility, parking and service access, all of which must be balanced with countervailing residential needs for privacy and seclusion. The feasibility of integral commercial needs to be driven by the retail demand, not by the nearby housing count.

However, one of the advantages to mixed-use buildings is the fact that once a commercial structure has been built, the space above often becomes "free space" and could easily be used for affordable housing such as rentals, generating additional income for the property. By free space, I mean that once the parking requirement and footprint of the commercial building have been accommodated, space above retail can be added cheaply because retail benefits alone should pay for the land and development costs. Further, retail and residential can have complementary parking requirements—housing needs more parking at night and retail needs more parking in the day.

Fig. 8.18 At *Two Worlds* in Pleasant Hill, CA, architect Donald MacDonald was able to combine ground floor commercial use with housing above. (Courtesy MacDonald Architects. Photo by Scott Zimmerman)

An example of this combination was built in Pleasant Hill, California. Designed by progressive San Francisco architect Donald MacDonald and called *Two Worlds*, the project combined 62 townhomes and 20,000 sq. ft of commercial space on a 4-acre suburban site. Both retail and professional users occupied the ground floor commercial, with about a quarter of the commercial space occupied by residents of the townhouses above.

Two Worlds provides a model for resolving the complexities of a mixed-use building—commercial fronting on the sidewalk, parking behind commercial and residential above. Pedestrian walkways and recreation occupy the center of the site. Trellis structures mitigate the views from the residential decks to the parking below.

Although there are other similar examples of integrated commercial and multi-family homes, they are the exception to the rule because they conflict with dogmatic zoning and administrative development practices. The growing popularity of village plans should make mixed-use multifamily buildings more common.

Future Multifamily Prototypes in New Communities

As stated earlier in the chapter, multifamily has typically been driven by lifestyle or affordable markets. For primary housing, multifamily has provided the first rung on the ladder of equity, or a step down in size for empty nesters at the other end of the housing cycle. It also provides a housing type for people with special needs and smaller households, and is a common prototype for senior/congregate housing.

These roles for multifamily will likely continue, but in much more integrated ways. Mixed-income communities, with mixed-density and perhaps even mixed-use buildings, will become increasingly popular as traveling and transportation become more expensive and time-consuming. Multifamily should return to models more like the historic urban setting of their origination with open space and commercial amenities proximate to their location.

Chapter 9
Manufactured Housing

I would like to briefly address manufactured and modular housing as a segment of new home production, particularly with affordable homes. For purposes of this discussion, I will only address manufactured housing that is substantially factory-built as modules, hauled by trailer to the site, and assembled with field labor. And in keeping with the scope of this book, the discussion will only address low-rise applications of manufactured housing, though still acknowledging that recently there have been tremendous inroads toward mid- and high-rise application of manufactured housing in urban settings.

Manufactured housing shipments peaked in the mid-1970s with around 575,000 units built, which at the time represented close to one-quarter of all housing starts. But for the most recent year, only approximately 64,000 manufactured units were shipped, or about 9 % of all housing starts.[1] This decline is anecdotally due to urbanization; manufactured housing is typically built in low-density settings on large rural lots. Only one-third of all shipments are built in communities, typically retirement villages where the land is leased.

It is curious that manufactured housing represents only a small portion of America's total housing stock, since under factory conditions, construction quality and costs can be controlled better for manufactured housing than for conventional field-built housing. According to the Manufactured Housing Institute, the per-square-foot cost of manufactured housing is about 50 % less than site-built

Fig. 9.1 Cost-effective manufactured housing has struggled to overcome the stigma from both buyers and builders that it was cheap and inevitably ugly. However, manufactured housing has a lot of potential to rival conventionally site-built housing, as seen in this manufactured home by Beracah Homes, Inc. in Delaware

© Springer International Publishing AG 2017
J. Wentling, *Designing a Place Called Home*,
DOI 10.1007/978-3-319-47917-0_9

housing. Yet both homebuilders and buyers have generally avoided manufactured housing though for different reasons.[2]

For buyers, manufactured housing carries an unjustified stigma of substandard construction. The public mistakenly associates manufactured housing with mobile homes, which in turn are considered low-class housing. Among buyers' legitimate concerns is the visual appearance of certain manufactured housing: box-like shapes and low-sloped roofs.

Homebuilders have avoided manufactured housing largely for a reason not unrelated to the above situation—it has a stigma with buyers. Why build something that doesn't sell? Still other reasons against manufactured housing are inability to customize plans for buyer revisions or options, and field problems that range from trade subcontractors to building inspectors, all of whom are wary of the unfamiliar. The housing industry is notorious for resisting change, and manufactured housing represents a big shift from the norm (site-built housing).

Manufactured housing is quite popular in some parts of America, particularly in rural locales where skilled labor is scarce. Several manufacturers have staked out specialties such as log construction and have become quite proficient in marketing those niche products. Others have focused on affordable housing and successfully competed with production builders. Yet manufactured housing has not reached its potential in design quality or production quantity in the USA when compared to other industrial countries—the most notable of which is currently Sweden.

Design Problems

There is much that can be done from a design perspective to improve the image of manufactured and modular housing. The following are some of the main design shortfalls:

1. *Shallow roof pitches*. Since manufactured homes need to be delivered to the site on a truck, and trailers need to fit under bridges, manufactured homes usually

Fig. 9.2 Modular or manufactured housing is generally weak in design: low-pitched roofs, random window and door placements, box-like exteriors, and poor detailing are common problems in the industry

have very shallow roofs. Roof pitches are generally in the 4/12 range, considered extremely low when compared to site-built homes. As discussed in Chap. 7, shallow roofs are associated with poor quality and cheap construction (although technically, that is unjustified).

2. *Random window and door placement.* Manufactured housing, when viewed from the outside, generally has no order or rhythm to window or door placements. Due to its common utilitarian role as affordable housing, establishing an exterior style has not been a design priority. Windows and doors are placed for the best interior location, without concern for the exterior appearance.

3. *Box-like plans and unbroken façades.* Because of the trailering requirement, manufactured homes are very long, yet shallow in depth. The most efficient way to design a home for maximum trailering dimensions is to create modules as wide and long as legally possible to fit on the back of a truck. This translates to modules being rectangles about 14 ft wide by 65 ft long. By attaching two of these modules together, the resulting house footprint would be 28 ft deep by 65 ft long. Triple wide models, which have recently become popular, would be as large as 36 by 80 ft. Such a monolithic structure then tends to forgo the design principles outlined in earlier chapters.

4. *Outdated building materials.* Manufactured housing builders do not always keep up with the rest of the industry in design trends. Components such as windows, doors, and finishes may be out of step with site-built market preferences. To save money, colors are often also poorly selected.

5. *Modular floor plans.* The interior floor plans of manufactured houses typically consist of smaller, defined rooms, as opposed to the more free-flowing spaces of conventional houses. This is because plans are based on combining modules that can be no more than 12–14 ft wide. The outside walls of these modules are structural, and manufactured housing builders do not like to put large openings in them. Hence, rooms are generally distinct and narrow.

Unit Design Opportunities

Most of the stereotypical negatives of manufactured housing could be overcome with the observance of just a few design priorities. Some specific points of improvement can make manufactured housing more visually appealing:

1. *Increase roof pitches.* On site-built housing, we recommend ranch designs to have roof pitches no less than 8/12, even on the most affordable of homes. There are ways of getting around the low roof pitches mandated by trailer limitations. In the shop, manufacturers can "hinge" the roof to achieve higher pitches. In this technique, roof trusses are put on hinges so that they can be lowered for transport, then extended to a higher pitch on site. Roof trusses could also be shipped to the site separate from the modules and field-applied with other veneers to allow for increased roof pitches (although this would increase costs).

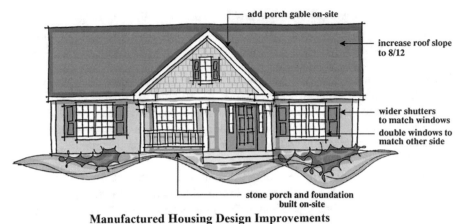

Manufactured Housing Design Improvements

Fig. 9.3 Manufactured housing aesthetics can be improved to appear more like well-done site-built houses by combining better design practices with field-built enhancements

2. *Improve window and door patterns.* Just as with conventional housing design, manufactured housing needs to represent a sense of order on exterior façades. This is a design adjustment that costs nothing to implement. Combined with field-applied elements that reduce the box-like appearance, manufactured housing can take on more qualities seen in conventional housing.
3. *Include site-built attachments such as porches and garages.* These field-applied attachments can, in conjunction with an improved window and door pattern, make a major difference in the overall image of manufactured houses. Porches, overframed front gables, and garages with breezeways are just some of the field-built structures that can be easily incorporated to finish off the design.
4. *Add window and door detailing.* As with conventional homes, doors and windows on manufactured housing should be "dressed up" with appropriate trim and accents such as shutters, muntins, and details. Manufactured builders tend to skimp on these features to cut costs.
5. *Apply some quality exterior veneers.* While the bulk of the manufactured housing is typically clad with siding, site-built alternatives may include masonry foundations, along with some veneer on façade areas. Chimneys are often good candidates for additional site-built structures that contribute quality veneers.
6. *Colors and other details.* Manufactured housing should be detailed with the same care taken as with site-built housing. Trim details, colors, etc. should reflect a design-conscious product. Outdated building components should be avoided.
7. *Use structural beams to open up modules.* In terms of interior design, manufactured housing needs to explore more fluidity in plans. As mentioned, rooms tend to be small and confining due to structural walls. These walls, however, can use beams in larger openings to create that feeling of spaciousness.

In our specific experience with a manufactured housing builder, we were able to apply many of these suggestions to a product line that the manufacturer was preparing to develop, to compete with production builders. Subsequently, at *Shady Grove Hills* in Shady Grove, Pennsylvania, extensive porches (that typically occupied at least one-third the overall length of the façade) were added and successfully mitigated the long boxy look of the base homes. A sense of rhythm was further instilled with the addition of a garage module, which connected to the main home with either a breezeway or an enclosed intermediate structure. The garage roof trusses were designed perpendicular to the main house trusses, providing even more visual interest to the façade.

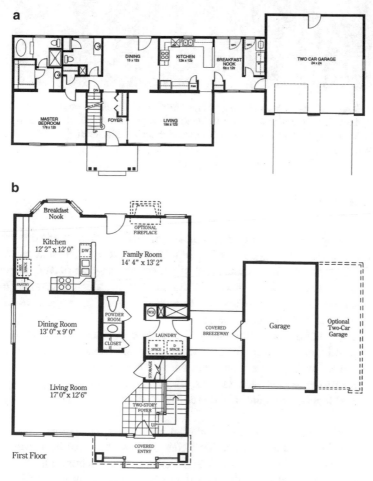

Fig. 9.4 (**a**) Modular plans typically suffer from compartmentalization in order to satisfy structural concerns at the lowest cost. (**b**) More open plans can be developed, as seen in a plan from *The Neighborhood at Fries Mill*, MJ. (Figure (**b**) courtesy Berkus Group)

Fig. 9.5 Homes at *The Neighborhood at Fries Mill* gracefully demonstrate that manufactured housing can overcome the stigma of poor design and be attractive as well when the design of modules and details are given the same level of attention that site-built housing is afforded. (See Fig. 9.5(**b**) for floor plan)

Community Design

In addressing the community design needs of manufactured home communities, it is helpful to know that manufactured houses can also be delivered as two-story models or Cape Cods. The common manufactured street scene of monotonous roof heights can be addressed by mixing different roof profiles on neighboring lots. Arranging models, such that Capes or two-stories break the pattern of one-level houses, improves community appearance.

Apart from varying building heights and roof profiles, manufactured house designs should look at turning the axes of neighboring roofs, separating the garage from the house, adding projections, and using other design techniques that create substantial differences between models. Typically, manufactured housing models tend to blend together in spite of the fact that they often actually possess different floor plans.

One successfully designed manufactured housing community is *The Neighborhood at Fries Mill* in southern New Jersey. Designed by Bill Devereaux (then of Berkus Group Architects of Washington, D.C.), this community overcomes the negatives associated with manufactured housing quite skillfully. *The Neighborhood* combines one- and two-story models with add-on garages to achieve a site variety usually not found in manufactured housing communities. The builders paid close attention to the details to create an Americana theme that was carried through in the interiors and marketing of the community.

Fig. 9.6 At *Schoolhouse Square*, site-built front porches and well-ordered façades worked lent themselves to providing an attractive street view for these affordable modular townhouses

Interiors at *The Neighborhood at Fries Mill* also exemplify how floor plans can be opened up to create more flowing spaces. Module walls seem very discreet and do not create closed-in spaces. *The Neighborhood* was a prototype community proposed for construction in several East Coast markets with variations in style to address local preferences and styles although that did not occur as planned.

In the multifamily category, *Schoolhouse Square* in Neptune, NJ is lauded for its design and as exemplary affordable housing. Designed by JKR Partners of Philadelphia, PA, this affordable townhouse community used attractive site-built front porches on entry façades, while parking was carefully placed at the rear of the units, achieving an attractive street view. Here, as with *The Neighborhood*, the designs show thoughtful window and door placements, material use and colors, and relation to the street. *Schoolhouse Square* used modular construction to level costs, so homes could be delivered under a local first-time buyer assistance program.

Modular construction is also ideal for meeting green building standards. LivingHomes, a company based in Santa Monica, CA, has been delivering affordable, LEED-certified houses as small as 1288 sq. ft to California markets. Using heavily insulated wall construction and green building materials, these houses are well suited for many of the smaller households seeking home ownership. The houses' flat roofs and modern styles further the savings for their buyers.

One sign that modular is currently ready to break into the mainstream US housing market is that the business conglomerate Berkshire Hathaway has been purchasing several manufactured housing companies throughout the country, which suggests they are working on a business plan to deliver entire communities using modular construction. This is only logical since this product and delivery system is ripe for expansion into the affordable housing market and will be successful when given added design attention by the industry.

Fig. 9.7 The LivingHomes company is combining sustainable design features with modern styling to deliver green, affordable, and attractive new modular homes to the Southern California market. (Images courtesy LivingHomes)

Notes

1. "2016 Quick Facts." Manufactured Housing Institute, Arlington, VA: 2016. www.manufacturedhousing.org
2. Ibid.

Chapter 10
Toward More Sustainable Homes and Communities

Perhaps the most significant factor to influence how new homes are built since the first edition was published has been the movement toward more energy efficient and sustainable design practices. Indeed, the concept of "going green" is everywhere, from cars to household appliances to the entire built environment. The topic of sustainable design is a book in and of itself, so in this chapter we will only touch on how green building practices are impacting the design and construction of new market-rate production housing.

There are two broad goals of sustainable design with respect to new homes: reducing consumption of resources and carbon emissions during the home's occupancy, and reducing the impact of initial construction of the house on the environment. These categories bear different weights by consumers since a reduction in resources used, such as water and energy, also means a reduction in utility bills. Consumers view this as a tangible reward for any added costs that green construction may incur, while protection of the environment is not. Surveys of potential homebuyers have consistently reinforced the position that consumers are in favor of protecting the environment through green building practices when they would only incur minimal additional costs to their new home purchase to do so.[1] This mentality is likely to change as younger households enter the housing market, but this is the overall sentiment of current buyers today.

Fig. 10.1 Photovoltaic solar panels that provide electric power are more frequently seen on the roofs of newer homes and communities as housing consumers seek lower utility bills and clean energy

© Springer International Publishing AG 2017
J. Wentling, *Designing a Place Called Home*,
DOI 10.1007/978-3-319-47917-0_10

Historical Background

One can note that colonial era settlers and Native Americans considered natural features, site orientation, and physical designs that would capture or deflect the sun's heat in their houses.

The pueblo settlements of the Southwest are a great example of this, as are the earth-covered structures of the Northern Plains. European settlers in the East used natural features to buffer winds and capture solar heating in their buildings as well. Most physical designs in the north were compact with few openings to retain heat inside the dwelling, with the Cape Cod and saltbox houses of New England having centrally located fireplaces and minimal exposed stretches of exterior wall and roof surfaces.

Fig. 10.2 (**a**) Traditional homes in the Southwest retain the benefits of thick adobe walls to reduce heat gain, as seen at this compound in Santa Fe, NM. (**b**) Homes in the coastal South had extensive covered balconies to capture breezes and cool the inside rooms during hot summer month as exemplified by the familiar "Charleston House" designs

Later American vernacular housing types that were responsive to local climates include the Tidewater folk houses with deep roof overhangs and porches in the Southeast, the French Colonial elevated designs of the Mississippi Delta, and the Spanish Colonial courtyard houses in the West. In the North, the thick masonry walls of colonial houses retained heat in the winter and maintained cooler indoor temperatures in the summer.

A major game-changer for housing design came in the late nineteenth century when fossil fuels began to replace wood as the predominant form of heating. Initially coal and oil, and later electricity and natural gas, provided the necessary energy for heating, cooking, and bathing inside the home. In the mid-twentieth century, air-conditioning for cooling was added to residential designs, and by the late 1970s it was considered a standard feature. With these developments, residential designs gradually became unhinged from the natural environment and focused inward on interior room arrangements and an insulated building envelope that would keep those rooms comfortable.

Earth Day, the EPA and the DOE

While conservationists such as Teddy Roosevelt and John Muir were advocates for preserving the natural environment earlier in the nineteenth century, the seminal moment for environmental protection in the twentieth century might be considered to be the first Earth Day on April 1, 1970. At this point, the degradation and pollution of the environment had become so severe that it became a national priority to establish laws to protect our waterways, air, and other natural resources.

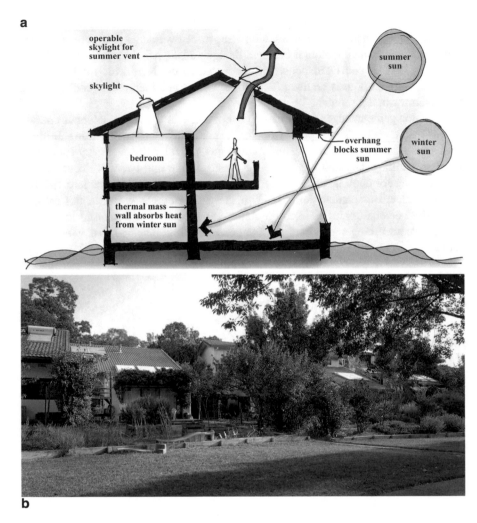

Fig. 10.3 (**a**) In the 1970s, passive solar design was a popular theme for some custom homes, which used site orientation and building design to absorb/deflect heat from the sun. At *Village Homes* in Davis, CA, an entire community was built with solar energy as a standard. (**b**). (Fig. (**b**) courtesy Kathy West Studios)

Therefore on December 2, 1970, Congress established the *Environmental Protection Agency* (EPA) to protect human health and the health of the environment.

In response to the Arab oil embargo of 1974, the *Department of Energy* was created in 1977 to formulate energy policy in the USA. Over time, the DOE began to provide minimum efficiency standards for residential appliances such as water heaters, furnaces, dishwashers, washers, and dryers. As we will discuss later, the DOE also created the popular *Energy Star* program, which addresses these appliance-based standards as well as entire home design standards.

In order to reduce pollution created by burning fossil fuels, early sustainable housing prototypes from the 1970s and later focused on renewable energy sources such as solar and wind as drivers of both community plans and individual house designs. Initially, solar panels were used to heat water as an energy source for a hot water supply and interior space heating. Later technologies introduced photovoltaic solar panels able to capture solar energy and convert it into electricity.

Although some state and local governments provided incentives for builders to experiment with adding renewable energy sources to market-rate housing, the cost-benefit did not justify solar panels to be added as a standard feature, or even as an option. It would not be until the early twenty-first century that the cost of solar panels had come down to the level where a reasonable payback period could allow homebuilders to offer them in the base price of homes.

In 2005, Congress passed the Energy Policy Act, which provided for a tax credit of 30 % of the cost to install solar electric systems and/or solar water systems. This legislation was later expanded to include other renewable energy installations and has made a significant impact on the use of solar panels for electricity generation, so much so that the Solar Energy Industries Association now predicts that solar capacity in the USA will triple by 2020 to power 20 million homes and represent 3.5 % of US electricity generation.[2]

The International Energy Conservation Code

Since adding renewable energy was not quite economically feasible in the last decades of the twentieth century, policymakers turned their attention to fortifying the building envelope to retain or deflect heat by ramping up insulation values given in the residential building codes. At the same time, the Department of Energy issued higher standards of efficiency in fossil fuel-burning household appliances.

The minimum standard that all residential buildings need to meet is the regionally adopted building code. Without meeting said standards, a building permit cannot be obtained. The *International Code Council*, or ICC, which is a US organization based outside of Chicago, IL, publishes a series of national building codes on a triennial basis. The body of codes published by the ICC includes the *International Residential Code* (IRC) and the *International Energy Conservation Code* (IECC), which are two of the codes that govern most single-family and townhouse construction. The ICC building codes are in turn adopted by the state legislatures with regional amendments, which will govern almost all construction in the state.

Building codes started including energy conservation standards around 1950 or so, in the form of minimum insulation requirements for exterior walls, ceilings, and floors based on the local climate. As energy became more costly following the Arab oil embargo of the 1970s, energy conservation efforts accelerated and the building codes responded by raising minimum standards.

The first International Energy Conservation Code was published in 1998 and was based on the previous Model Energy Code of 1995. The primary objective of the code with respect to housing is increasing resistance to heat loss though the building envelope—which comprises the walls, ceiling, windows, and doors of a single structure. In addition, the code includes standards for mechanical systems and lighting. With each triennial edition of the code, these standards are increased to add additional resistance to heat loss in the building envelope and developed with some input from builder organizations so as to balance added energy efficiency with the added costs that can challenge housing affordability.

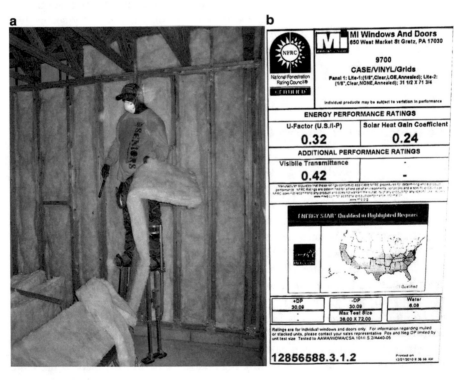

Fig. 10.4 (**a**) Building codes focused on the building envelope as well as mechanical equipment efficiencies to reduce energy consumption in new homes. (**b**) Insulation values for walls and windows were increased with each edition of the residential building code to reduce summer heat gain and winter heat loss

Green Rating Programs

Most people have seen the terms "LEED-certified" or "Energy Star-rated," in reference to two of the many green rating programs used to provide consumers with third-party verification that buildings or products are in fact better for the environment. These are perhaps the best-known programs though others include the "Green Globes" and the "National Green Building Standard." Regional organizations such as the Southface Energy Institute in Atlanta have their own programs such as "Earthcraft." The international *Passivhaus* standard originated in Germany and now has a following in many countries, including the USA (where it is known as *Passive House*).

In addition, most regional power companies have incentive programs for energy conservation that, if implemented, lead to reduced pricing for the home's utilities or one-time financial rebates. Some homebuilders also have their own green rating programs, which they may brand and market for their new homes to give them an edge over the competition.

These rating programs have varied goals and requirements to be met for certification. Some programs such as Energy Star are primarily focused on energy conservation, while the LEED programs have a more holistic approach and encompass planning, construction, and material use in addition to energy consumption. As might be expected, the more stringent and complex programs are more costly to implement, and therefore they have less of a following in market-rate housing. Builder surveys indicate Energy Star is by far the most common green rating program while LEED is the least common.[3]

A common argument that design practitioners had early on with rating programs was whether the impact of a home's size, location, and setting should be considered in the rating. For example, a 1500 sq. ft house on a quarter acre lot near public

Fig. 10.5 Builders advertise the energy efficiency of their new homes with various rating programs as consumers place a high value on increased savings on their future utility bills

transportation and services is going to be greener, in terms of lifecycle impact, than a 5000 sq. ft house in a remote, large lot location several miles from the nearest services. For some time, such factors were not considered in the holistic rating programs but they were incorporated in later releases, mainly in the form of losing rating points for excessive size and remote lot location. That is no longer the case.

The Energy Star® Program

The Energy Star program is an international voluntary standard for energy efficient consumer products. Created in 1992 by the EPA and the DOE, the Energy Star label indicated that the branded consumer appliances generally used 20–30 % less energy than required by federal standards. The program started with computer products and in 1995 was expanded to cover residential heating and cooling systems in new homes. Currently there are more than 40,000 Energy Star products including residential appliances, lighting, home electronics, and more. As of 2006, about 12 % of new housing starts in the USA were Energy Star qualified.[4]

Fig. 10.6 This new home constructed by progressive builder Saussy Burbank and 505 Design in Charlotte, NC is an Energy Star rated home, as are all the homes built by Energy Star Partner builders. (Image courtesy Saussy Burbank. Photography by Zan Maddox)

In order to qualify for an Energy Star rating, a new home must consume at least 15 % less energy than standard designs that meet the building code minimum. This typically requires higher insulation values in the building's walls, roof, windows, and doors. In conjunction, the mechanical systems that provide heating and cooling are designed to higher standards.

Since Energy Star has the most name recognition with consumers—from seeing the label on appliances and computer equipment—it is the most popular rating system for potential homebuyers. In addition, many power companies offer discounts on utility bills for Energy Star rated homes. Energy Star is also popular with builders, since the third-party rating process is relatively streamlined, as opposed to more rigorous green certification programs.

Unlike other green certifications, Energy Star historically did not address sustainability issues outside of energy consumption such as water conservation, square footage of the house, landscaping, or other issues typically considered important in sustainable housing design. In the latest release of the program, however, these issues are becoming included in the checklist for achieving the rating.

The HERS® Index

Many green rating programs such as Energy Star rely on the Home Energy Rating System, or *HERS Index*, to determine just how much more or less efficient a new home is, compared to a home built to minimum code standards. A standard house has a HERS Index of 100, while a house that consumes 30 % less energy than the standard house has a rating of 70. A house that consumes 30 % more than a standard house has a rating of 130.

The HERS rating program is administered by the Residential Energy Services Network, or *RESNET*, which is a nonprofit organization founded in 1995 to help homeowners reduce their utility bills by making their homes more energy efficient. RESNET certifies independent contractors for performing the rating tests and maintains a national database on houses that have been rated. Buyers can therefore verify the rating of a house prior to purchase, similar to the third-party records available for automobiles and other major items of purchase.

LEED

LEED stands for Leadership in Energy and Environmental Design and is administered by the U.S. Green Building Council (USGBC), a nonprofit organization based in Washington, D.C. Established in 1993, their mission was to promote sustainability in the building and construction industry. Currently, the organization consists of tens of thousands of member organizations and chapters representing many professionals including builders, manufacturers, educators, and nonprofit sectors.

The USGBC provides certifications for buildings at various levels (Certified, Silver, Gold, and Platinum), based on complex rating systems that take into account energy consumption, sourcing and manufacturing processes of materials, methods of construction, and impact on the natural environment. The organization also provides a variety of credentials for design professionals based on a comprehensive testing and work experience program.

The initial rating system was not based on building occupancy or use, but as the organization evolved, rating programs for buildings as well as the corresponding certifications for designers became segmented accordingly. For the homebuilding industry, *LEED for Homes* has been developed to rate single-family houses as well as low-rise attached housing up to six stories. *LEED-Neighborhood Development* provides a rating system for community design, which takes into account an entire collection of housing units and the landscaping, streets, and utilities around them to determine what level of sustainability has been achieved.

Unlike Energy Star, the LEED for Homes certification depends on much more than reducing energy consumption. To determine the sustainability of a home, virtually every aspect of construction is taken into account: site access to public transportation, conservation of water usage, selection of materials used, sourcing and location of said materials, methods of removing waste from the site, and levels of indoor air quality are but a few of these aspects.

In this respect, LEED for Homes is a much more rigorous certification to achieve than Energy Star, and although the LEED brand is highly recognized by consumers, builders of market-rate single-family housing have not embraced LEED for Homes, due to the cost of getting certified by a third party. However, for larger multifamily buildings, the LEED certification is more common, as the cost can be amortized over a larger construction budget.

National Green Building Standard™

The National Green Building Standard (NGBS) is a rating program developed by the National Association of Home Builders (NAHB), in conjunction with the ICC, to be a more builder-friendly rating system as an alternative to LEED. The NGBS is a further developed version of NAHB's *Model Green Building Guidelines*, initiated in 1998. Work on the NGBS began in 2004 and the first edition was released in 2008. In January 2009, the American National Standards Institute (ANSI) approved the program as ICC 700–2008—the first residential green building standard to receive ANSI approval.

Similar to LEED, the NGBS awards points in various categories to determine the level of the rating—bronze, silver, gold, and emerald—and considers site preparation, resources used in construction, energy and water conservation, indoor air quality, construction waste disposal, and other life cycle concerns. The NGBS applies to single-family, multifamily, and residential remodeling projects. As with LEED, it also requires third-party verification of the requirements, but builders find the NGBS to be a less cumbersome standard than LEED.

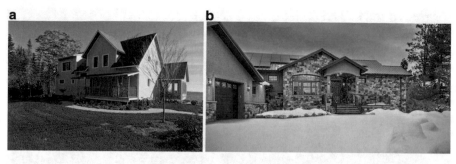

Fig. 10.7 Both (**a**) Passive House and (**b**) the Net Zero Ready Homes rely on heavily insulated envelopes—the skin of buildings—to prevent heat loss (or heat gain) so as to reduce energy consumption to a bare minimum. (Fig. (**a**) courtesy Ecocor High Performance Buildings. Fig. (**b**) courtesy Mantell-Hecathorn Builders, photography by Marona Photography)

Passive House

The *Passivhaus* (or *Passive House*) movement began in the colder climates of central Europe (hence the German namesake) and focuses on energy efficiency, primarily through super-insulation and air-tightness of the building exterior envelope. The first Passivhaus residences were built in Darmstadt, Germany in 1990, and in 1996 the Passivhaus Institut was established in Darmstadt as well. As of 2016, there are estimated to be more than 25,000 Passive House buildings throughout the world.

Here in the USA, the Passive House Institute US, or *PHIUS*, was founded in 2002 in Urbana, IL. This organization has provided training for designers and builders, as well as established partnerships with major energy providers and other agencies to promote and expand the Passive House brand. It also holds an annual national conference and is a clearinghouse for information on building products and energy consumption monitoring tools. They have trained more than 1700 designers, builders, and other design practitioners, and maintain a directory of over 120 single-family and multifamily projects that have been certified.

Passive House standards do focus on energy consumption, as measured by energy units per square foot, and following these standards equates to something close to a 90 % reduction in energy use over the space of a typical building. The Passive House brand is gaining traction, primarily with custom homes, but also with government-sponsored multifamily designs. The mainstay of the program is a higher insulated building envelope, and is being applied to other high-performance designs such as Net Zero Energy Ready Homes.

Net Zero Energy Ready Homes

This program was developed by the Department of Energy to encourage builders to provide houses that could, at some point, be converted to zero-energy homes. Houses would have increased insulation and more energy efficient windows

Fig. 10.8 The design of Net Zero Energy Ready Homes is similar to Passive House design in that a heavily insulated building envelope reduces the need for mechanical heating and cooling. Photovoltaic solar panels are anticipated to be added to the home at a future date (if not with initial construction) to provide the energy needed to power a minimally sized mechanical system

installed as standard, with the potential for solar panels to be installed at a later date. This allows homebuilders' prices to be more competitive with basic code-compliant models offered for sale in the same market, while allowing buyers to be able to add solar panels upon purchase or later on, based on their budgets.

Typically, Zero Energy Ready Homes (ZERH) are modeled for energy consumption based on projected annual utility costs *without* solar panels versus projected annual utility costs *with* solar panels. Projected annual energy cost savings is also analyzed for a similar code-compliant house that does not have the increased insulation and higher performance windows and doors of ZERH.

In addition to the higher insulation values, the ZERH program requires designs to meet the standards established by the Energy Star program as well as indoor air quality and water conservation standards. Homes certified for the ZERH program are rated by the Home Energy Rating Service, which provides a HERS index as discussed above. (Thus, a ZERH is the product of the various pieces of green building programs enacted by the DOE.) According to the *Net Zero Energy Coalition*, there are over 6000 units across North America that produce as much renewable energy as they consume, or could do so with slight modifications. This is expected to increase by six-fold over the next few years, as the program gains traction with homebuyers.[5]

Healthy Homes

The *Healthy Homes Program* is not so much a rating system but an advocacy program undertaken by federal agencies such as HUD (Department of Housing and Urban Development) and the CDC (Center for Disease Control) as well as independent nonprofits and state and local governments. Healthy Homes is focused on a variety of issues ranging from indoor air quality to safety hazards such as lead poisoning and pesticides. Although there is no healthy home rating program as of yet, HUD does offer grant money for demonstration projects and acts as a clearinghouse for information.

The *Green and Healthy Homes Initiative* (which is different from Healthy Homes) is one example of a nonprofit that promotes both characteristics of green design and health issues. The eight attributes of Green and Healthy Homes are dry, pest-free, clean, contaminant-free, safe, well-maintained, well-ventilated, and energy efficient. This program grew out of the *Coalition to End Childhood Lead Poisoning* and expanded into a wider scope of health issues.

Builders can always market aspects of energy efficiency, sustainable construction, and healthy home issues on their own by explaining to consumers how their homes are different by addressing issues discussed above. This can be more cost effective than adhering to rating programs and is a popular way to appeal to consumer preferences for green and healthy housing.

Resilient Design and Safe Homes

In addition to the goals of green and healthy home programs, we should note that other movements encouraging resilient and safe housing design should be part of the sustainability discussion. Natural disasters such as floods, hurricanes, wildfires, earthquakes, and tornadoes devastate parts of the country each year which result in a tremendous loss of life, property damage, and community destruction. Resilient housing design anticipates some of these extreme weather events and incorporates defensive measures into construction practices to limit or reduce the damages typically associated with these events.

Fig. 10.9 Resilient design anticipates how periodic forces of nature can damage or destroy homes. For protection from flooding, homes are to be raised above government-mapped minimum floor elevations

One program that addresses resilient design is the *Federal Alliance for Safe Homes* (FLASH) which is a Tallahassee, FL based nonprofit which offers education, training, and support materials to assist builders and the public with constructing more fortified, resilient housing. These techniques are based on the extreme weather events posed to various regions. In general, the most common threat to most of the country is wind events, which occur as hurricanes along the coast and tornadoes in the heartland. Working with building material suppliers, construction to resist wind forces, as well as lateral movements from earthquakes are a large part of the recommended practices encouraged. In addition, tornadoes can have such high wind forces that small basement "safe rooms" under the home are recommended for these areas since it is often impossible to resist tornado force winds.

Other recommended building practices can limit damages from coastal or river flooding, wildfires, hail, lightning strikes, and heavy snowfalls. Damage from these weather events and disasters can be mitigated with various construction techniques that will allow the structure to survive relatively intact to avoid costly and wasteful re-building.

Another organization that promotes resilient design is the *Institute for Business and Home Safety* (IBHS), another nonprofit funded by the insurance industry. This Tampa, FL based organization has similar goals to promote resilient design. IBHS funds and conducts research into how various fortified designs can resist extreme weather events and publishes the findings in order to promote design practices to reduce the loss of life and property damage.

Smaller Homes Are Green Homes

Sometimes we overlook the most basic solutions to complex problems. By simply cutting down on interior square footage, many of the goals of building sustainable communities can be achieved. Unfortunately, the average size of new homes in the USA only continues to grow year after year, while average household size continues to decline. In the 1950s, the average square footage of a new home was under 1000 sq. ft when the average household size was 3.37 persons; fast-forward to 2010 and the average house was now at 2392 sq. ft while the household size declined to 2.59 persons. Car ownership followed suit and grew from an average of 1.43 cars per household in 1950 to 2.11 in 2010.[6] The overall picture is one of greater material and energy consumption waste.

In the early 1990s, architect Sarah Susanka embarked on a new career as an author and published The Not-So-Big House, which immediately struck a chord with many buyer profiles—particularly empty nesters looking to downsize in their next home. The premise of the book, and later editions, was that smaller homes of higher quality make a lot of sense on a number of issues.

Fig. 10.10 A pocket neighborhood of smaller houses clustered around a common green is a popular design concept for infill parcels in already-established neighborhoods. *Conover Commons* in Redmond, WA, designed by Ross Chapin and developed by The Cottage Company, embraces the concept in order to promote interactions between neighbors, while achieving a BuiltGreen 4-Star Rating and Energy Star certification. (Courtesy Ross Chapin Architects)

Detached home or multifamily unit buyers are more likely to reduce their energy consumption and carbon footprint simply by residing in smaller living spaces.

This is particularly true when buyers are moving from older, poorly insulated houses into newly constructed smaller homes built to the current energy code. This change is likely to reduce energy/resource consumption by half of what was consumed in older homes. Multifamily designs with two or more common walls further reduce heat loss, by reducing the surface area exterior wall through which to lose heat. Smaller houses occupy less land area since they can also be built at higher densities. A great example of this is the *pocket neighborhood*, where small homes are built close together in a mews-like fashion around a central green. Langley, WA architect Ross Chapin has popularized this concept for developing passed-over infill parcels of land in housing markets nationwide.

Green Rated Homes Are Growing

Of the green rating programs discussed above, Energy Star has the largest market share by a wide margin. This is partly due to longevity and public awareness of the Energy Star brand though purchases of electronics and appliances. The program has over 20 years of rating new home construction and claims 1.6 million certified homes built to date. This represents $4.7 billion in energy savings on utility bills. All of the other rating programs are also growing, particularly the Net Zero Ready Homes and Passive House programs.

Settlement Patterns: Are Suburbs Unsustainable?

Many people in the planning and design community are of the opinion that suburban housing communities are generally unsustainable, and should therefore be discouraged though land use policies. The argument is that dispersed and low-density housing go hand in hand with higher automobile emissions, unnecessary land consumption, additional stormwater runoff, and oversized dwellings. Meanwhile, cities are touted for offering compact living spaces, access to mass transit, and services within walking distance.

For every opinion piece or research paper that argues against suburban living, there are others that state the exact opposite. Cities still have drawbacks when it comes to sustainability: increased emissions due to traffic congestion, reduced access to solar energy, higher amounts of energy used to access and condition vertical buildings, and less carbon-absorbing trees and foliage. Most urban housing is smaller in square footage, but considering smaller household size, the square footage per occupant is comparable to that of suburban homes.

It has been said that New York City, one of the highest density cities in the USA, is also the "greenest" due to the availability and use of mass transit, walking, and bicycling to get around as opposed to the use of automobiles in lower density cities.

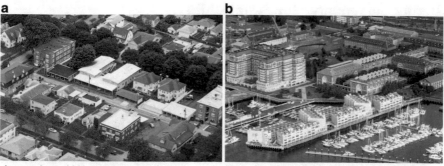

Longview, WA 27.8 units / acre   Charlestown, MA 26.7 units / acre

Fig. 10.11 A common opinion is that low-density settings are unsustainable and should be discouraged, while city living is far more efficient in terms of energy and land consumption. However, here we see (**a**) a neighborhood with three-story walk-ups can achieve a density similar to an urban area with high rises (**b**). (Photos courtesy Alex S. MacLean/Landslide Aerial Photography)

However, when considering other livability concerns such as cost of living, quality of schools, air quality, and access to affordable housing and higher paying jobs, cities can be seen in a negative light. While New York can be considered the greenest city in the country, it has also been labeled the most unequal in terms of the divide between socioeconomic classes.

Affordable housing is perhaps the most significant issue that differentiates dense urban areas from suburban communities. The cost to build a housing unit in an urban setting is approximately three times that for a suburban location. Typically, higher density housing will utilize concrete and steel construction as opposed to less expensive wood frame. Other elements of higher density housing such as elevators, multiple stairs, and more sophisticated mechanical systems are also required by code. This results in much higher costs for ownership housing, and even urban rental housing is typically prohibitively high.

Fig. 10.12 (**a**) *New Port* is a compact, affordable community in Portsmouth, VA, developed on an old Navy housing site. (**b**) A nature preserve around a stormwater retention basin serves as a park, with trails running throughout the community. Public transportation and other public amenities are nearby as well

Rating Communities for Sustainability

As with individual housing units, there are a number of systems for rating the green aspects of entire neighborhoods, towns, or regions. Early examples of lists that define sustainable communities were produced by federal, state, and local government agencies, as well as professional organizations such as the American Planning Association, or APA, and the Congress for New Urbanism, or CNU. These rating systems are diverse as well as numerous; however, most can be considered under the umbrella of "smart growth" policies for more sustainable living patterns.

Smart Growth

Smart Growth is a movement that emerged in the 1980s to promote more efficient patterns of urban expansion other than low-density settlements that consumed large tracts of land, otherwise known as "sprawl." Simply put, smart growth encourages new development to occur within the service area of existing utilities, roads, and services as opposed to extending these services into new areas. Common terms for the former are "brownfields" vs. "greenfields" for the latter, with greenfield development being seen as less desirable from a sustainability point of view.

The various smart growth programs throughout the country are quite numerous and diverse in their mission. Most of the organizations are coalition members of the Washington DC based nonprofit Smart Growth America and have missions that range from land conservation to encouraging bicycle and walking trails.

While smart growth concepts generally apply to urban growth patterns on a larger scale, for new residential communities land may become available through redevelopment agencies that is ideal for new home communities. These parcels are typically already serviced by roads and utilities and also have access to public transit. Local governments often make the land available but also impose design and price standards since public funds are involved with transferring the land to private builders.

LEED-Neighborhood Development: LEED-ND

The LEED for Homes rating program, discussed previously, at one point was heavily criticized for not considering the building location in the rating. Therefore a LEED for Homes rated house could be built in a remote location where long-distance auto travel (or even flying) would be necessary to access the home. As a result, the LEED-Neighborhood Development, or LEED-ND, program was introduced to assess how sustainable dwellings within a community setting can be.

Fig. 10.13 *High Point* in West Seattle, WA is a neighborhood development project by Mithun architects, whose design adhered to stringent green building and land use standards to provide healthy affordable homes for the Seattle Housing Authority. (Photography © Doug Scott)

The LEED-ND program was developed as a joint effort of the Natural Resources Defense Council, the Congress for the New Urbanism, and the U.S. Green Building Council, which administers the program. In addition to protecting the environment through advocacy for natural resource protection and action to reduce climate change, it also includes social concerns such as affordable housing, public health, and social equity. Some of the issues considered in the physical design include the variety of dwelling types, the schools and playgrounds within walking distance, and the stores and commercial uses within the community.

Walk Score® and Walk-Ups

Walk Score is an online platform that provides a "walkable rating" for any home in the USA simply by inputting the street address. Developed by entrepreneur and Harvard grad Mike Mathieu and now owned by RedFin, a real estate consulting firm, Walk Score locates nearby features such as restaurants, grocery stores, shopping, convenience stores, parks, schools, and cultural amenities and provides a score from 1 to 100. Scores above 50 are considered "walkable," while lower than 50 are "car dependent" in varying degrees.

The Walk Score site measures walkability with a patented system based on a 30-min walking radius with points awarded in 5-min increments, meaning amenities within 5 min are given more points than those further away than 5 min on a declining

Fig. 10.14 Walk Score® rates the *walkability* of every home, based on surrounding public amenities and the proximity of these amenities to each home. Walkability has become a valuable metric for homebuyers

basis within the 30-min radius. The Walk Score site is quite popular with consumers looking for a home in a walkable community setting. However, in a recent RedFin poll of over 750 homebuyers, walkability was rated as important by just 18 % of buyers, while school quality was 40 % and yard or green space was 34 %.[7]

A leading proponent of walkable communities is Brooking Institute Fellow Christopher Leinberger, who has written a series of books and publications on the walkability of various metro areas such as Boston and Washington, DC. In these publications, he includes a rating system for ranking walkable neighborhoods based on certain characteristics. Regions rank as Copper, Silver, Gold, or Platinum, and the criteria address both economic and social equity considerations.[8]

Transit-Oriented Development

As previously discussed, Transit-Oriented Developments (TODs) are new or revitalized communities that are located with easy access to public transportation, thus reducing automobile dependency and carbon emissions while providing improved access to jobs and economic opportunities to car-less individuals. Legislation at the federal, state, and local levels has created TOD programs to provide financial incentives for housing and mixed-use buildings proposed to be located near transit stops.[9]

Typically, transit-oriented developments tend to be higher density, multifamily, with a mixed-use component needed to support the transit use—features which are generally outside the scope of this book. Some lower density communities can participate in these programs simply by being located near local bus routes or in infill parcels that are within walking distance of other mass-transit options.

Sustainable Homes and Communities

Clearly the movement toward more sustainable residential designs and communities is gaining traction. As little as ten years ago, many builders saw green building programs as a passing "fad" that will go away in time. However the challenge still remains is to match green building practice costs with consumer attitudes regarding the value of living in a green certified home. As mentioned in the beginning of the chapter, buyers are able to justify added costs for energy savings, but are tepid about

Fig. 10.15 This light rail station in Charlotte, NC connects the downtown to suburban communities, thereby relieving traffic congestion and reducing carbon emissions from automobile usage

other added costs for sustainable features that do not have a payback period associated with it.

Often higher end clients and custom homebuyers are the ones to pursue sustainable program labels, as they are more likely to have the funds for the extra costs and see the label as important to their commitment to the environment. Publicly financed affordable housing programs also tend to be more likely to undertake green rating program design since it raises their stature as a responsible use of public money and suggests reduced operational costs for the future occupants.

Also mentioned earlier, the attitudes of the private sector buyer vary by age range, with a much higher commitment to sustainable processes being shown by younger buyers. This should propel sustainable design into practice in greater and greater percentages of the market. This, in addition to an increasingly stringent energy code that determines the baseline efficiencies of all homes, will continue to raise the expectation of sustainable and energy efficient design to become the new normal as we move forward.

Notes

1. Quint, Rose. "Survey Says Home Trends and Buyer Preferences". *National Association of Homebuilders*: 2014..
2. "Why Will Solar Win?" *BUILDER Magazine*: March 2016.
3. "Green Building Survey." *Professional Builder Magazine*: Sept 2007.
4. "Build Energy Star® Qualified Homes With REC." *Energy Star Builders Brochure*. Energy Star Program.
5. "Net-Zero Homes Gaining Traction." *Professional Builder Magazine*: March 2016.
6. Davis, Stacy C., Susan W. Diegel & Robert G. Bouncy. "Transportation Energy Data Book: Edition 31." Office of Energy Efficiency and Renewable Energy, U.S. Department of Energy: July 2012
7. Starace, Alex. "Homebuyers Don't Think Housing Is An Important Political Issue. We Disagree." Published on *RedFin Real Time*: 15 March 2016.
8. Leinberger, Christopher and Michael Rodriguez. "Foot Traffic Ahead, Ranking Walkable Urbanism in America's Largest Metros." *George Washington University*, Washington, DC: June 2016.
9. Davis, Stacy C., Susan W. Diegel & Robert G. Bouncy. "Transportation Energy Data Book: Edition 31". Office of Energy Efficiency and Renewable Energy, U.S. Department of Energy: July 2012.

Conclusion

What I have attempted to cover in this book are design issues related to the most common types of housing and communities being built in the USA today—and outline some ideas on how to make them more livable, sustainable, and enjoyable for the occupant—without adding a lot of cost. For that reason, we have focused on low-to-moderate density housing types. In addition, we have focused on housing types of modest size and cost in the interest of seeing new communities accommodate households from varied income levels.

Although this book is about suburban housing and communities, it is in no way intended to be an argument against city living. My home city of Philadelphia is currently enjoying a wonderful urban renaissance—gaining population, jobs, and new businesses after many years of decline. I have lived in several neighborhoods in the city and my office has been located in the heart of the city since inception. The power of urbanism to bring diverse people together in communal settings such as parks, restaurants, and cultural venues is attractive to people of all ages.

Fig. A.1 (a) Single family detached houses were built by the Philadelphia Housing Authority to replace substandard rowhome housing in central Philadelphia, and are similar to suburban homes seen in the city's outlying areas such as this home in *Eagleview* in Chester County, PA (b)

© Springer International Publishing AG 2017
J. Wentling, *Designing a Place Called Home*,
DOI 10.1007/978-3-319-47917-0

a   **b**

Fig. A.2 (**a**) New multifamily housing construction in urban areas of Philadelphia is not that dissimilar from multifamily housing built outside the city. (**b**) This suburban community in Hampton, VA also calls for three-story walk-up "stacked townhouse" buildings with street level private entrances

While city living is desirable for some, others prefer the less dense and more private environment of suburban or even exurban housing and communities. What I find interesting is that some redeveloped areas of cities are becoming less dense, while many new suburban communities are becoming more dense. In Philadelphia, once you get beyond the central business district of high-rise offices and residential towers, the housing is not all that different from typical multifamily housing in the suburbs. In fact, some blighted areas have been redeveloped with single-family detached homes. Most often, however, the new construction in cities is some variation of townhouses or walk-up multifamily.

By contrast, in new suburban communities the trend is moving toward higher density housing along the themes discussed in this book, in favor of mixed-density, mixed-use walkable communities with access to open space and other amenities and services. In a sense, there seems to be a universal interest in moderate density living. Some market researchers have dubbed this "new urban-lite," which attempts to provide some of the diversity and vitality found is urban areas while retaining some of the privacy and open space that suburban communities typically offer.

The challenge for both settings is building affordable housing. Philadelphia, like many legacy cities, is somewhat of a "tale of two cities" with almost all new market-rate housing being in the luxury category, while the new affordable housing is either subsidized or government provided—and is nowhere close to meeting demand. Many of the new jobs are in the service industry and the average household income simply does not allow families to buy or rent new market-rate housing types.

Traditionally, and as recently as the first edition of this book, suburban communities could offer lower density affordable housing as an alternative to substandard older homes in impoverished urban neighborhoods, allowing households to move to a better home and community. These options have been dwindling since the first

Fig. A.3 This diagram effectively illustrates how moderate density housing types can bridge the gap between single family detached and higher density elevator buildings. (Illustration © Opticos Design, Inc. Reprinted with permission)

edition twenty years ago. Since that time the cost to build new housing has skyrocketed due to a variety of forces. Approvals have become more cumbersome, significant impact fees are commonly assessed, and other regulatory barriers have been put in place, all of which has driven up the cost of delivering a new home in suburban settings to beyond the reach of entry-level buyers. Currently, the national average for the costs to bring a residential lot to market is 25 % of the total cost of the house as we noted in Chap. 2.

These same forces have also impacted the costs of building urban housing, and here the cost increases are even more severe. As mentioned in Chap. 10, urban housing can be up to three times more costly per square foot than in outlying areas. Increasing the supply of affordable and workforce housing is an extremely complex discussion that involves government, politics, finance, and socioeconomic implications. However, suffice to say, we need governments at every level to step up and provide support to developers attempting to deliver moderate cost housing.

For either venue, moderate density "walk-ups" seem to make the best case for affordable housing. One movement that seems to articulate this idea is the "Missing Middle Housing" argument, advanced by Dan Parolek of Opticos Design in Berkeley, CA. This suggests that land use regulations need to be modified to encourage some of the housing prototypes more commonly built in the first half of the twentieth century and include duplexes, triplexes, courtyard buildings, townhouses, and three-story apartments of wood frame construction. In addition, "accessory units" such as a living unit over a garage or a very small detached dwelling on the same lot would be permitted. These housing types would bridge the gap between single family and mid-rise elevator buildings located in the core of the city.

This idea makes total sense for increasing our affordable housing stock. Here, in Philadelphia, Habitat for Humanity and other nonprofit housing developers focus on building duplexes and townhouses in neighborhoods which once may have had high-rise public housing. Now returning vacated land to the original rowhome pattern and moderate density seems to make more sense than constructing elevator buildings.

By definition, a suburb cannot exist without a city that it grew from. One can also say that while cities have been spreading out since Roman times, and recently cities have been shrinking in population and density. What is interesting is how similar housing in cities and suburbs is becoming. Hopefully, our governments will increase efforts to make affordable housing a part of that transition in both cities and suburbs, so neighborhoods in both settings are more diverse, vibrant, and livable.

Other Books on Residential Design

Alexander, Christopher. 1977. A Pattern Language. New York: Oxford University Press. A fasci-
nating manifesto based on the theory that designs for cities, as well as for individual homes can
be organized with a collection of time-honored "patterns".

Alexander, Christopher. 1979. A Timeless Way of Building. New York: Oxford University Press.

Alexander, Christopher. 1985. The Production of Houses. New York: Oxford University Press.
This is an application of the theories in A Pattern Language. Alexander was hired to construct
a village in Mexicali, Mexico. This book traces the sequencing of local participation in the
design and construction process.

Anthenat, Kathy Smith. 1991. American Tree Houses and Playhouses. Whitehall, VA: Betterway
Publications, Inc. Color illustrations of playhouses for kids. Now available from www.books-
forcomfort.com, 2007.

Appleyard, Donald. 1981. Livable Streets. Berkeley and Los Angeles, CA: University of California
Press. Includes behavioral studies of neighborhoods with different street designs. Also has
ideas to control traffic through residential areas.

Arendt, Randall. 2015. Rural By Design: Planning for Town and Country, Second Edition.
Chicago, IL and Washington, DC: American Planning Association Planners Press. An encyclopedia of community planning approaches with case studies.

Arendt, Randall. 1999. Crossroads, Hamlet, Village, Town. Chicago, IL and Washington, DC:
American Planning Association Planning Advisory Service Report Number 487/488.

Babcock, Richard F. 1966. The Zoning Game: Municipal Practices and Policies. Cambridge, MA:
Lincoln Institute of Land Policy. An early identification of poor planning practices by local
governments.

Babcock, Richard F. 1985. The Zoning Game Revisited. Cambridge, MA: Lincoln Institute of
Land Policy. An update of the classic original edition.

Baldassare, Mark. 1988. Trouble in Paradise: The Suburban Transformation of America.
New York: Columbia University Press.

Bender, Richard. 1973. A Crack in the Rear View Mirror: A View of Industrialized Building.
New York: Van Nostrand Reinhold.

Benevolo, Leonardo. 1971. The Origins of Modern Town Planning. Cambridge, MA: MIT Press.
Translated from Italian, this book looks at the European experience with town planning, primarily in the 19th and 20th centuries.

Berg, Donald J. 2005. American Country Building Design: Rediscovered Plans for 19th Century
Farmhouses, Cottages, Landscapes, Barns, Carriage Houses and Outbuildings. New York:
Sterling Publishing Company.

Berger, Bennett M. 1960. Working Class Suburbs: A Study of Auto Workers in Suburbia Berkeley,
CA: University of California Press.

© Springer International Publishing AG 2017
J. Wentling, *Designing a Place Called Home*,
DOI 10.1007/978-3-319-47917-0

Birmingham, Stephen. 1978. The Golden Dream: Suburbia in the 1970s. New York: Harper and Row.

Bruegmann, Robert. 2007. Sprawl: A Compact History. Chicago: University of Chicago Press.

Calthorpe, Peter. 1993. The Next American Metropolis: Ecology, Community and the American Dream. New York: Princeton Architectural Press.

Calthorpe, Peter, Doug Kelbaugh et al. 1989. The Pedestrian Pocket Book: A New Suburban Design Strategy. New York: Princeton Architectural Press. Essays and drawings developed at a design charette held at the University of Washington in 1988. The Pedestrian Pocket is a small 50 to 100 acre model community that balances housing with jobs, linked by mass-transit to a larger metropolitan area.

Calthorpe, Peter and William Fulton. 2001. The Regional City. Washington, DC: Island Press.

Campoli, Julie and Alex S. MacLean. 2007. Visualizing Density. Cambridge, MA: Lincoln Institute of Land Policy.

Campoli, Julie. 2012. Made for Walking, Density and Neighborhood Form. Cambridge, MA: Lincoln Institute of Land Policy.

Chamberlin, S. and J. Pollock. 1983. Fences, Gates and Walls: How to Design and Build. Los Angeles, CA: HP Books - Price Stern Sloan, Inc.

Chermayeff, Serge, and Christopher Alexander. 1963. Community and Privacy: Toward a New Architecture of Humanism. Garden City, NY: Doubleday.

Clark, Clifford Edward Jr. 1986. The American Family Home 1800-1960. Chapel Hill, NC: University of North Carolina Press.

Clawson, Marion and Peter Hall. 1973. Planning and Urban Growth: An Anglo-American Comparison. Baltimore, MD: John Hopkins University Press.

Corn, Joseph J. (ed.). 1986. Imaging Tomorrow: History, Technology and the American Future. Cambridge, MA: MIT Press.

Crawford, Margaret. 1995. Building the Workingman's Paradise. New York, NY: Verso. Profiles new communities sponsored by manufacturers, employing well-known planners.

Davies, Thomas D. Jr. and Kim A. Beasley. 1992. Fair Housing Design Guide for Accessibility. Washington, DC: National Association of Homebuilders.

Deck Planner: 25 Outstanding Decks You Can Build. Tuscon, AZ: Home Planners, Inc. 1990.

Dietz, Albert G.H. 1974. Dwelling House Construction. Cambridge, MA: MIT Press. Originally published in 1948, this classic explains the nomenclature and technical information regarding housing construction.

Downing, Andrew Jackson. 1969. The Architecture of Country Houses. New York: Dover Publications, Inc. Originally published in 1850 as a pattern book for builders, this edition includes extensive text describing Jackson's residential design philosophy.

Downing, Andrew Jackson. 1981. Victorian Cottage Residences. New York: Dover Publications, Inc. Another pattern book with details.

Duany, Andres, Jeff Speck and Michael Lydon. 2010. The Smart Growth Manual. New York: McGraw-Hill, Inc.

Early American Home Plans (120). Farmington Hills, MI: Home Planners, Inc. 1984. A modern day pattern book by a large stock plan service. Includes text describing how historical styles can be adapted to contemporary programs.

Easterling, Keller. 1993. American Town Plans. New York: Princeton Architectural Press. With computer-generated graphics, this book traces the history of town planning.

Eichler, Ned. 1982. The Merchant Builders. Cambridge, MA: MIT Press.

Ewing, Reid. 1991. Developing Successful New Communities. Washington, DC: The Urban Land Institute. On case study overview of conventional new communities.

Fader, Steven. 2000. Density by Design: New Directions in Residential Development, Second Edition. Washington, D.C.: Urban Land Institute. An update of the 1988 original by edited by the author of this book, James Wentling along with Lloyd Bookout.

Fishman, Robert. 1982. Urban Utopias in the Twentieth Century: Ebenezer Howard, Frank Lloyd Wright, Le Corbusier. Cambridge: MIT Press.

Fishman, Robert. 1987. Bourgeois Utopias. New York: Basic Books. A superb history of Anglo-Saxon suburbs.

Fleming, Ronald L. and Lauri A. Halderman. 1982. On Common Ground: Caring for Shared Land from Town Common to Urban Park. Harvard, MA: The Harvard Common Press, Inc. A historical look at the evolution of commonly owned and maintained open space, with contemporary guidelines for design.

Flint, Anthony. 2006. This Land, The Battle over Sprawl and the Future of America. Baltimore, MD: The Johns Hopkins University Press. An excellent chronicle of the current state of Smart Growth and the New Urbanist Movement.

Foley, Mary Mix. 1980. The American House. New York: Harper and Row.

Gans, Herbert J. 1967. The Levittowners: Ways of Life and Politics in a New Suburban Community. New York: Pantheon.

Garreau, Joel. 1991. Edge City, Life on the New Frontier. New York: Doubleday. A travelogue-like look at the impact of decentralized cities on human values.

Gehl, Jan. 1987. Life Between Buildings. New York: Van Nostrand Reinhold. Ideas to make the public realm more human scale and social.

Gutman, Robert. 1985. The Design of American Housing. New York: Publishing Center for Cultural Resources. This book explains who is designing our homes and how it is done.

Gowans, Alan. 1986. The Comfortable House, North American Suburban Architecture 1890-1930. Cambridge, MA: MIT Press. A scholarly review of the initial suburban building boom. From 1890 - 1930 more homes were built than in the nation's entire history, in styles that included Bungalow, Saltbox, Shingle, Tudor and Gothic. These were "comfortable" houses.

Hayden, Dolores. 1984. Redesigning the American Dream. New York: Norton. An argument for smaller homes for smaller households, many of which are single women or mothers.

Herbert, Gilbert. 1984. The Dream of the Factory-Made House: Walter Gropius and Konrad Wachsmann. Cambridge, MA: MIT Press.

Hirschman, Jessica Elin. 1993. Porches and Sunrooms. New York: Michael Friedman Publishing Group. A color idea book.

Hirt, Sonia A. 2014. Zoned in the USA, The Origins and Implications of American Land-Use Regulation. Ithaca, NY: Cornell University Press.

Hunter, Christine. 1999. Ranches, Rowhouses and Railroad Flats: American Homes and How They Shape Our Landscapes and Neighborhoods. New York: W.W. Norton and Company.

Jackson, Kenneth T. 1985. Crabgrass Frontier: The Suburbanization of the United States. New York: Oxford University Press. A well-researched and thoughtful history of suburbanization in the United States.

Jacobson, Max, Murray Silverstein, Barbara Winslow. 1990. The Good House: Contrast as a Design Tool. Newton, CT: The Taunton Press. A collection of design approaches to satisfy more human and emotional needs in residential architecture, written by the co-authors of A Pattern Language.

Jakle, John A., Robert W. Bastian, Douglas K. Meyer. 1989. Common Houses in America's Small Towns: The Atlantic Seaboard to Mississippi Valley. Athens, GA: University of Georgia Press.

Jandl, H. Ward. 1991. Yesterday's Houses of Tomorrow: Innovative Homes 1850 to 1950. Washington, DC: Preservation Press. A case study look at homes that introduced new ideas about housing.

Jones, Robert T. 1929 (1987). Authentic Small Houses of the Twenties. New York: Dover.

Kemp, Jim. 1990. American Vernacular. Washington, DC: AIA Press. Examines over 50 indigenous styles of American design.

Kendig, Lane. 1980. Performance Zoning. Washington, DC: Planners Press. A wonderful argument for basing development controls on performance standards such as floor area ratios and impervious cover- instead of conventional controls such as minimum lot sizes, setbacks, etc.

King, Anthony D. 1984. The Bungalow, The Production of a Global Culture. London: Butterworth.

Klaus, Susan L. 2002. A Modern Arcadia, Fredrick Law Olmsted Jr. and the Plan for Forest Hills Gardens. Amherst and Boston: University of Massachusetts Press.

Kone, D. Linda. Land Development, 10th Edition. 2006. Washington, DC: BuilderBooks, National Association of Homebuilders. An overview of the technical challenges in the land development process.

Kostof, Spiro. 1987. America by Design. New York: Oxford University Press.

Krieger, Alex with William Lennertz. 1991. Towns and Townmaking Principles. Cambridge, MA: Harvard Graduate School of Design. Based on a study of the Duany/Plater-Zyberk team and their work at Seaside.

Krier, Rob. 1983. Elements of Architecture. London: Architectural Design Publications. A typological catalog of the elements of architecture.

Kulash, Walter. 2001. Residential Streets. Washington, DC: Urban Land Institute, National Association of Home Builders, American Society of Civil Engineers, Institute of Transportation Planners. A look at how streets can be built with more respect for human scale and the environment. Includes practical engineering information on reasonable roadways.

Lancaster, Clay. 1985. The American Bungalow, 1880-1930. New York: Abbeville Press. Traces the origins of the beloved bungalow, which served as a prototype affordable home for generations.

Langdon, Philip. 1987. American Houses. New York: Stewart, Tabori & Chang. A rich color pictorial with insightful commentary on all types of housing.

Leinberger, Christopher. 2012. The Walk-Up Wake-Up Call: Washington, DC. Washington, DC: The George Washington School of Business, Center for Real Estate and Urban Analysis. Download at GW Business website, includes walkable rating levels for economics and social equity considerations. Several other metros were also studied by Mr. Leinberger, a fellow at the Brookings Institute.

MacDonald, Donald. 2015. Democratic Architecture. International: ORO Editions.

McAlester, Virginia, and Lee McAlester. 1984 (Revised 2015). A Field Guide to American Houses. New York, NY: Alfred A. Knopf, Inc. A comprehensive guide to historical residential architecture in America.

Moore, Charles, Gerald Allen, Donlyn Lyndon. 1974. The Place of Houses. New York: Holt, Rinehart and Winston. A classic and poetic look at the residential design process.

Moore, Charles, et al. 1983. Home Sweet Home: American Domestic Vernacular Architecture. New York: Rizzoli International Publications.

Mohney, David and Keller Easterling, editors. 1991. Seaside: Making a Town in America. New York: Princeton Architectural Press. Provides a history of the town of Seaside, zoning and building codes, and over 120 of the buildings constructed.

Mohr, Merilyn. 1988. Home Playgrounds. Scarborough, Ontario: Camden House/Firefly Books.

Naisbitt, John. 1982. Megatrends. New York: Warner Books. A bestselling book that reviews the impact of technology and information on our society.

Naisbitt, John. 1990. Megatrends 2000. New York: Avon Books. More trends to be considered in this update.

Nivola, Pietro S. 1999. Laws of the Landscape, How Policies Shape Cities in Europe and America. Washington, DC: Brookings Institute Press.

Nolon, John R. and Duo Dickenson. 1990. Common Walls/Private Homes. New York: McGraw-Hill, Inc. Case studies of successful multi-family prototypes with commentary.

"Not In My Backyard": Removing Barriers to Affordable Housing. Washington, DC: Department of Housing and Urban Development. 1991. A look at the real reasons for the lack of affordable housing in America- local government and over-regulation.

Outdoor Playhouses and Toys. New York: Sterling Publishing. 1986.

Pearson, David. 1989. The Natural House Book: Creating a healthy, harmonious and ecologically sound home environment. New York: Fireside/Simon and Schuster. An excellent book on healthy residential design issues.

Putnam, Robert D. 2000. Bowling Alone, The Collapse and Revival of American Community. New York, NY: Simon & Schuster.

Prowler, Donald. 1985. Modest Mansions. Emmaus, PA: Rodale Press. A practical guide to designing smaller, more modest individual homes.

Questions and Answers about Building: Fine Homebuilding. Newton, CT: The Taunton Press, Inc. 1989. Published by the same group that produces "Fine Homebuilding" magazine, this book addresses technical details of building homes.

Ramsey, Dan. 1992. Fences, Decks and other Backyard projects. Blue Ridge Summit, PA: TAB Books (McGraw-Hill).

Rapaport, Richard. 1992. Your Future Home: Architect Designed Houses of the Early 1920s. Washington, DC: AIA Press.

Reiff, Daniel D. 2000. Houses from Books, Treatises, Pattern Books and Catalogs in American Architecture 1738-1950: A History and Guide University Park, PA: The Pennsylvania State University Press.

Rogers, Kate Ellen. 1962. The Modern House, USA: Its Design and Decoration. New York: Harper and Brothers.

Rowe, Peter G. and John Michael Desmond. 1986. The Shape and Appearance of the Modern Single Family House. Cambridge, MA: Joint Center for Housing Studies of MIT and Harvard Universities.

Rowe, Peter G. 1991. Making A Middle Landscape. Cambridge, MA: MIT Press. A well-researched, impartial look at suburban metropolitan developments. Written by the dean of Harvard's Graduate School of Design.

Rybczynski, Witold. 1986. Home, A Short History of an Idea. New York: Viking Penguin, Inc. The idea is comfort- this book is a study of comfort as an ideal in house design.

Rybczynski, Witold. 1989. The Most Beautiful House in the World. New York: Viking. The most beautiful house in the world is the house one builds for oneself. This is a book about that process inspired by the author's own experience.

Schmitz, Adrienne. 2004. Residential Development Handbook. Washington, DC: The Urban Land Institute. A comprehensive guide for residential development.

Schuttner, Scoot. 1993. Building and Designing Decks. Newtown, CT: Taunton Press, Inc.

Seamon, David Ed. 1993. Dwelling, Seeing and Designing. Albany, NY: State of New York Press.

Sedan, Paul S. 1992. The Factory-Crafted House: New Visions of Affordable Home Design. Old Saybrook, CT: The Globe Pequot Press.

Sergeant, John. 1984. Frank Lloyd Wright's Usonian houses: Designs for Moderate Cost One-Family Homes. New York: Whitney Library of Design.

Sexton, Richard. 1995. Parallel Utopias, The Quest for Community. San Francisco, CA: Chronicle Books. Essays of Seaside, FL and Sea Ranch, CA.

Sherwood, Gerald E. and Robert C. Stroh. 1988. Wood Frame House Construction. Armonk, NY: Armonk Press. Available through the NAHB Bookstore, this book is an excellent technical guide to residential construction details.

Solomon, Daniel. 1993. ReBuilding. New York: Princeton Architectural Press. San Francisco-based architect and planner's views of suburbia with recommended solutions.

Stein, Clarence. 1957 and 1966. Toward New Towns for America. Cambridge, MA: MIT Press.

Stern, Robert A.M. & David Fishman, Jacob Tilove. 2013. Paradise Planned, The Garden Suburb and the Modern City. New York, NY: Monacelli Press.

Stern, Robert A.M. 1986. Pride of Place, Building the American Dream. Boston, MA: Houghton Mifflin Company. Companion to the PBS TV series, this lavishly illustrated book addresses housing and community design along with other building types.

Stern, Robert A.M. 1981. The Anglo-American Suburb. London: Architectural Design Publications, Ltd. This book summarizes prototypical suburbs in England and America starting in the early 19th century.

Sternlieb, George. 1986. Patterns of Development. Rutgers, NJ: Center for Urban Policy Research. A good explanation of how the national housing market is impacted by demographics, migration, and economic cycles.

Stevenson, Katherine Cole and H. Ward Jandl. 1986. Houses by Mail. Washington, DC: The Preservation Press. A listing of home designs that were available through Sears, Roebuck and Company from 1908 to 1940.

Stiles, David. 1993. Sheds: The Do-it Yourself Guide for Backyard Builders. Charlotte, VT: Camden House Publishing. Includes some great ideas for "accessory structures."

Strombeck, Janet A. and Richard H. Strombeck. Backyard Structures: Designs and Plans. Delafiled, WI: SunDesigns.

Susanka, Sarah & Kira Oblensky. The Not So Big House. 2008. Newton, CT: The Taunton Press, Inc. This is the tenth anniversary edition of the original book, which was followed by a series of "Not-So-Big" titles along similar themes of downsizing our living spaces.

The Backyard Builder's Book of Outdoor Building Projects. Emanus, PA: Rodale Press, Inc. 1987

Turner, John F.C. 1976. Housing by People: Toward Autonomy in Building Environments. London: Marion Boyars Publishers Ltd. A study which makes a case for low density housing looking at international markets.

Van Buren, Maurie. 1991. House Styles at a Glance. Atlanta, GA: Longstreet Press. A guidebook to identify historical housing styles.

Van der Ryn, Sim, and Peter Calthorpe. 1986. Sustainable Communities. San Francisco, CA: Sierra Club Books. Along with other contributors, the authors describe environmentally responsible community design issues in the urban and suburban context.

Venturi, Robert, Denise Scott Brown, and Steven Izenour. 1972. Learning from Las Vegas. Cambridge, MA: MIT Press.

Venturi Scott Brown & Associates, On Houses and Housing. New York: St. Martins Press. 1992. A look at how a respected design firm approaches residential commissions.

Wachs, Martin and Margaret Crawford, editors. 1992. The Car and the City: The Automobile, The Built Environment and Daily Urban Life, Ann Arbor, MI: University of Michigan Press.

Wallis, Allen. 1991. Wheel Estate. New York: Oxford University Press.

Walker, Lester. 1997. American Shelter, An Illustrated Encyclopedia of the American Home, Revised Edition. Woodstock, NY: The Overlook Press. The author explains the history of American housing with detailed sketches and text. A well researched, comprehensive, unique reference book.

Watkins, A.M. 1988. The Complete Guide to Factory Made Houses. Chicago, IL: Longman Financial Services Publishing, Inc.

Weaver, Gerald L. 1993. Fireplace Designs. Cincinnati, Ohio: Betterway Books.

Weiss, Marc A. 1987. The Rise of the Community Builders. New York: Columbia University Press. A scholarly review of the transition in American housing production from small individual builders to large scale community developers.

Williams, Norman., Edmund Kellogg and Peter Lavigne. 1987. Vermont Townscape. Rutgers, NJ: Center for Urban Policy Research. Provides insight for attaining what has come to be a model for many new communities.

Wentling, James W. and Lloyd W. Bookout. 1988. Density by Design. Washington, D.C.: Urban Land Institute.

Wentling, James W. 1995 Housing by Lifestyle, The Component Method of Residential Design. New York, NY: McGraw-Hill.

Wright, Gwendolyn. 1981. Building the Dream: A Social History of Housing in America. Cambridge, MA: MIT Press. A unique view of the domestic environment in America, from the New England town through industrial villages to suburban sprawl and public housing.

Wolf, Peter. 1981. Land in America, Its Value, Use and Control. New York: Pantheon. An excellent overview of how the American built environment got to where it is today.

Wolfe, Tom. 1981. From Bauhaus to Our House. New York: Farrar Straus Giroux. A witty indictment of modernism.

Yaro, Robert D., Randall G. Arendt, Harry L. Dodson, and Elizabeth A. Brabec. 1988. Dealing with Change in the Connecticut River Valley: A Design Manual for Conservation and Development. Cambridge, MA: Lincoln Institute of Land Policy. An award-winning design manual with compelling sketches and drawings illustrating the benefits of cluster development.

Index

© Springer International Publishing AG 2017
J. Wentling, *Designing a Place Called Home*,
DOI 10.1007/978-3-319-47917-0

Printed in the United States
By Bookmasters